AF348601

Agathe Keller

Expounding the Mathematical Seed

Volume 2: The Supplements
A Translation of Bhāskara I on the Mathematical Chapter
of the Āryabhatīya

Birkhäuser Verlag
Basel · Boston · Berlin

Author

Agathe Keller
Rehseis CNRS
Centre Javelot
2 place Jussieu
75251 Paris Cedex 05
e-mail: kellera@paris7.jussieu.fr

A CIP catalogue record for this book is available from the Library of Congress,
Washington D.C., USA

Bibliographic information published by Die Deutsche Bibliothek
Die Deutsche Bibliothek lists this publication in the Deutsche Nationalbiographie;
detailed bibliographic data is available in the internet at http://dnb.ddb.de

ISBN 978-3-7643-7292-7 Birkhäuser Verlag, Basel – Boston – Berlin

© 2006 Birkhäuser Verlag, P.O.Box 133, CH-4010 Basel, Switzerland
Part of Springer Science+Business Media
Cover design: Micha Lotrovsky, CH-4106 Therwil, Switzerland
Cover illustration: The cover illustration is a free representation of
"Rat and Hawk" (made by Mukesh)
Printed on acid-free paper produced from chlorine-free pulp. TCF ∞

Vol. 2/SN 31:   ISBN 978-3-7643-7292-7   ISBN: 978-7643-7593-5 (eBook

9 8 7 6 5 4 3 2 1

Contents

Appendix: Some elements of Indian astronomy   **186**

Glossary   **197**

Bibliography   **227**

Index   **235**

List of Figures

List of Tables

How to read this book?

This volume contains the supplements for the translation presented in the first volume. The supplements aren't made to be read alone.

Indeed, Volume I contains an English translation of a VIIth Century Sanskrit commentary written by an astronomer called Bhāskara, and an extensive Introduction to the text. Because Bhāskara's text alone is difficult to understand, I have added for each verse commentary a supplement which discusses the linguistic and mathematical matter exposed by the commentator. These supplements are gathered in the present volume (Volume II), which also contains glossaries and the bibliography. The two volumes should be read simultaneously.

Abbreviations and Symbols

When referring to parts of the treatise, the *Āryabhaṭīya*, we will use the abbreviation: 'Ab'. A first number will indicate the chapter referred to, and a second the verse number; the letters 'abcd' refer to each quarter of the verse. For example, 'Ab. 2. 6. cd' means the two last quarters of verse 6 in the second chapter of the *Āryabhaṭīya*.

With the same numbering system, BAB refers to Bhāskara's commentary. Mbh and Lbh, refer respectively to the *Māhabhāskarīya* and the *Laghubhāskarīya*, two treatise written by the commentator, Bhāskara.

[] refer to the editor's additions;

⟨⟩ indicates the translator's addition;

() provide elements given for the sake of clarity. This includes the transliteration of Sanskrit words.

Supplements

The first part of Bhāskara's commentary on the mathemathematical chapter of the *Āryabhaṭīya* (e.g. his introduction to the chapter and the two first verse commentaries) has not been given any supplements. However, explanatory footnotes with references to secondary literature have been provided with the translations.

A   BAB.2.3

A.1   Arithmetical squaring and its geometrical interpretation

In answer to an ambiguous objection[1]:

> *āyatacaturaśrakṣetrādiṣu vargakarmaṇo 'stitvāt teṣām*
> *asamacaturaśrāṇām api vargasaṃjñāprasaṅgaḥ|*
> ⟨Objection⟩ Because a square operation exists in rectangular fields, and
> so on, there is the possibility for the name 'square' to be ⟨given to⟩ fields
> which are not equi-quadrilaterals also.

Bhāskara prescribes the construction of a square made by the diagonals of four
rectangles[2]. This diagram, as seen in Figure 1, supposedly "shows" that the arithmetical squaring of the length of a diagonal corresponds to the area of a square.

Figure 1: Bhāskara's diagram

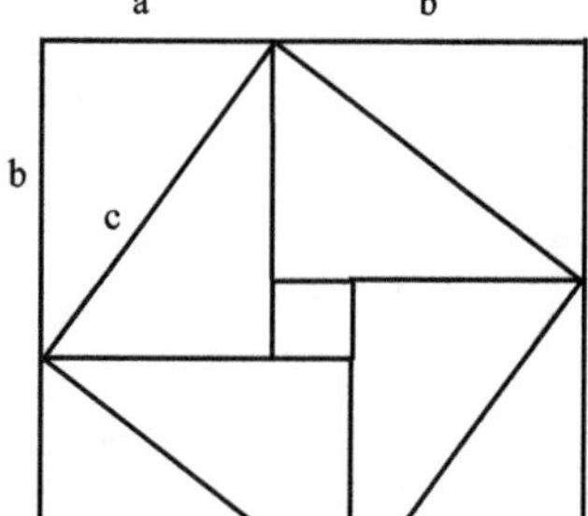

Several difficulties arise concerning this objection and the following paragraph.

First of all, the objection concerns the action of naming "square" (*varga*) fields
that wouldn't even be equilateral quadrilaterals. Bhāskara does not answer directly
on this point[3].

Secondly, an expression used by Bhāskara when describing the construction of this
field remains open to several interpretations. The description starts in this way[4]:

> *samacaturaśrakṣetram ālikhya aṣṭadhā vibhajya* ...
> When one has sketched an equi-quadrilateral field and divided ⟨it⟩ in
> eight ...

---

[1][Shukla 1976; p. 48, lines 9-10]

[2][Shukla 1976; p. 48, lines10-16]

[3]We can notice, however, that even if he states before that the object *samacaturaśra* (equi-
quadrilateral) has the name *varga* (square), he in fact never uses the latter for a geometrical
object. A *varga* in his commentary is always the result of the arithmetical operation of squaring.

[4][Shukla 1976; p. 48, line 10]

The question is then: how should one understand the expression
"*aṣ/dtadhā vibhajya*" ?

Implicitly, as can be seen in Figure 2, the editor considers that the square constructed has sides that measure 8.

Figure 2: Bhāskara's diagram in Shukla's edition

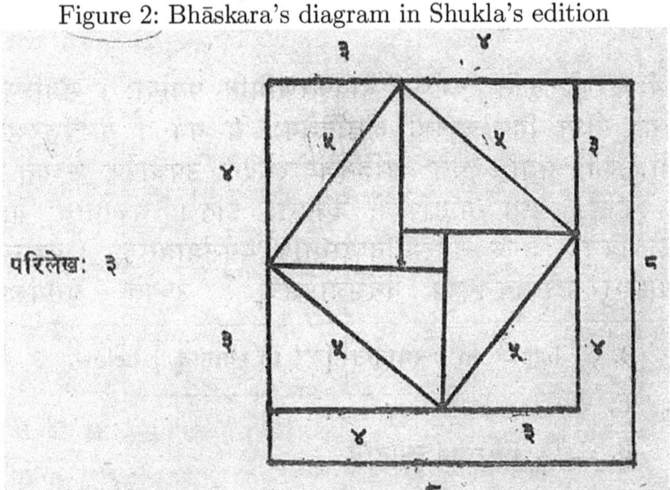

The understanding of the expression *aṣṭadhā vibhajya* (cut into eight) would then be that the rectangles are drawn by cutting into the sides of the squares. However the diagram that can be seen in our photographic copy of mss D, does not show such a square. This may be seen in Figure 3.

Figure 3: Bhāskara's diagram in a manuscript

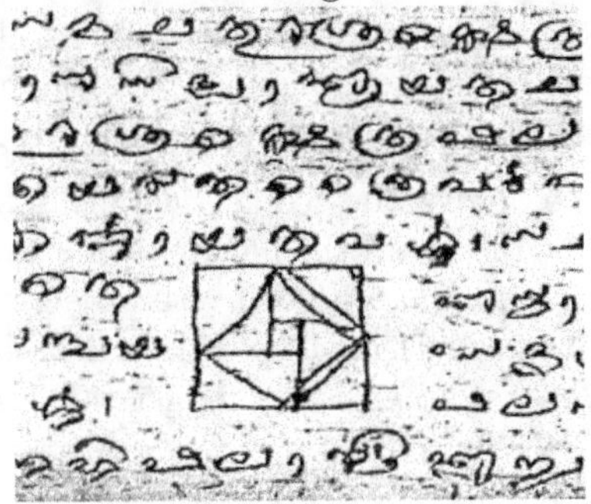

Another understanding of the expression could be to count the sub-surfaces, cut into the square whose sides measure 7, by the four rectangles and their diagonals. This is illustrated in Figure 4.

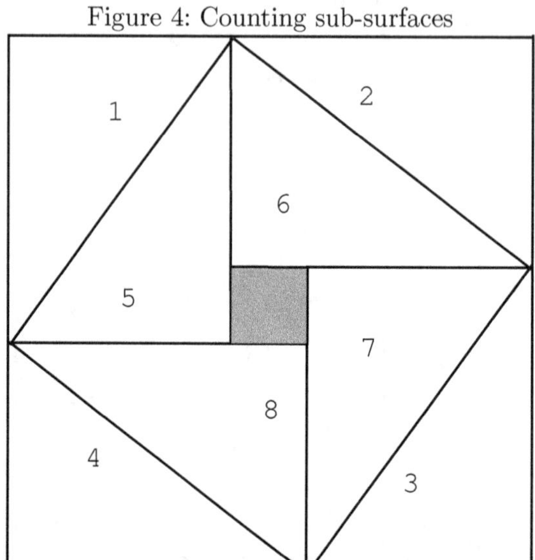

Figure 4: Counting sub-surfaces

Indeed these cuts draw eight right-angle triangles. The square in the middle would be left out because it is not considered in Bhāskara's reasoning. However, because one needs to omit the innermost square, this interpretation remains unsatisfactory.

Finally, one can consider that once the square whose sides measure 7 is constructed, the four rectangles and their diagonals are drawn in eight strokes. These strokes are illustrated in Figure 5.

None of these alternative interpretations prevents the expression from remaining quite enigmatic.

Returning to the problems occurring in the paragraph at stake we can note that the meaning of the objection remains ambiguous. We do not know what is a '*vargakarman*' (square-operation): Is it the numerical squaring of any length?

Certainly, Bhāskara's goal is to discuss the geometrical meaning of the squaring of a length, as when previously he discussed the nature of the *karaṇī* operation[5]. We believe that the expression *vargakarman* used in the objection does not concern the squaring of any length, but only that of a diagonal or hypotenuse (*karṇa*). Neither the questioner nor Bhāskara mentions the fact that this could be true for any length.

Indeed, it is surprising that his answer to the objection does not concern the arithmetical square of the side of any geometrical figure. His first reply runs as follows:

---

[5]See the introduction to the *gaṇitapāda*.

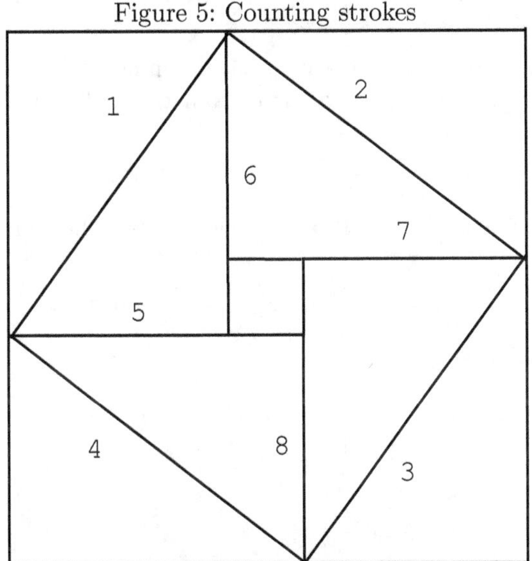

Figure 5: Counting strokes

naiṣa doṣaḥ| teṣu api yo vargaḥ sa samacaturaśrakṣetraphalam|
This is not wrong. In these ⟨fields⟩ too, a square is the area of an equi-
quadrilateral field.

The demonstrative (*teṣu*) refers to a list of fields given in the objection (rectangles,
etc.). Bhāskara's drawing illustrates the squaring of the diagonals of a rectangle.
He adds, referring certainly to a right-angle triangle:

> *tribhuje 'py etad eva darśanam, ardhāyatacaturaśratvāt tribhujasya|*
> Just this exposition (*darśana*) ⟨exists⟩ in a trilateral (*tribhuja*) also,
> because a trilateral is half a rectangle.

Even though this discussion does not concern directly the "Pythagoras Theorem"[6]
it is closely related to it.

Let us look at Figure 1 page 2 again. The area of the square in the middle can be
seen as the square of the diagonal of the rectangle (c^2). But we can also consider
the area of the first drawn square. This is equal to the square of the sum of the two
adjoining sides of the rectangle ($(a+b)^2$). Now if we cut off the areas of the four
triangles that corner this big square, we obtain once again the area of the square
in the middle. The area of each triangle is half the area of one rectangle ($4 \times \frac{ab}{2}$)
"because a trilateral is half a rectangle". So we then see that $c^2 = (a+b)^2 - 2 \times ab$.

---

[6]Quotation marks are used to indicate that the name is a convention with a story to it, and
that we do not consider that Pythagoras is the real discoverer of this property of right-triangles.

From which the formulation of the "Pythagoras Theorem" (stated in Ab.2.17.ab), algebraically $c^2 = a^2 + b^2$, may be deduced.

Even though Bhāskara does not elaborate this reasoning, it is noteworthy that the diagram he describes can be used in a geometrical demonstration of the "Pythagoras Theorem".

Figure 6: Ganeṣa's 'proof' of the 'Pythagoras Theorem'

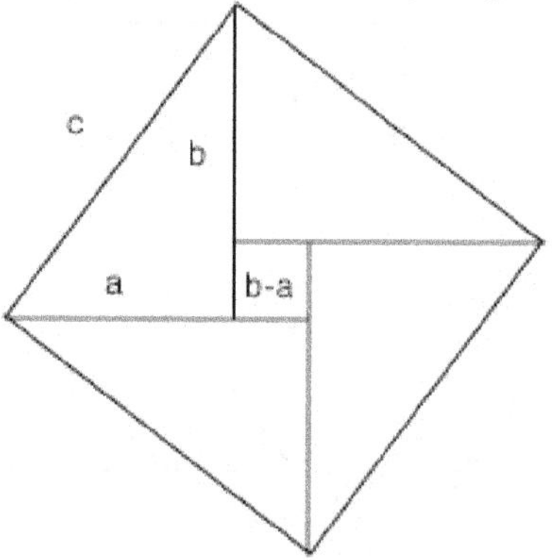

One can note that the "Pythagoras Theorem" was known and used by the authors of the *śulba-sūtras*, who considered it always in a rectangle. Ab.2.17.ab. as interpreted by Bhāskara, on the other hand, is almost systematically used in reference to a right-angle triangle. Concerning such a type of field before the time of Bhāskara, Datta & Singh are of the opinion that it was known by Āpastamba who used it in a proof of the "Pythagoras Theorem"[7]. However, no such field appears in any of these two authors' works. Its existence is deduced by Datta & Singh through the fact that its properties are used by Āpastamba and Baudhyāna, in the procedure for enlarging squares.

A similar type of field is known to have been presented for proofs or verifications of the "Pythagoras Theorem" after the time of Bhāskara I, by Bhāskara II[8] and by some of his commentators (namely Ganeṣa)[9]. But only the triangular part is considered with different lengths. This is illustrated in Figure 6.

In this diagram, the area of the interior small square whose sides are equal to $b - a$ (so the area is $(b - a)^2$) is increased by the area of the four triangles whose sides are a and b (the area of each triangle is therefore $\frac{ab}{2}$). This gives the area of the big square whose sides are the hypotenuse of the four triangles (in other words: $c^2 = (a - b)^2 + 4(\frac{ab}{2}) = a^2 + b^2$). This last reasoning uses also the fact, mentioned by Bhāskara I, that 'a trilateral is half a rectangle'.

---

[7][Datta&Singh 1980; p.134-135]

[8][Jain 1995; p.57]

[9][Srinivas 1990;p.39]

A.2 Squares and cubes of greater numbers

A.2.1 Squaring

Bhāskara quotes a rule (included in Shukla's list of quotations from other works [Shukla 1976; Appendix V, p.347]) to square numbers with more than one digit.

> *antyapadasya ca vargaṃ kṛtvā dviguṇaṃ tad eva cāntyapadam|*
> *śeṣapadair āhanyād utsāryotsārya vargavidhau||*
> When one has made the square of the last term, one should multiply
>   twice that very last term|
> ⟨separately⟩ by the remaining terms, shifting again and again, in the
>   operation for squares||

The procedure is elliptic for we do not know how it was carried out practically. How were the successive computations set down? Where did the final square appear? And some expressions are ambiguous. Indeed, the statement 'shifting again and again' (*utsārya utsārya*) can have a double meaning. It may refer to the successive multiplications of the doubled last term with the following digits, or to the repetition of the process itself, considering one after the other the digits of the number to be squared. Even though we have considered, in the reconstruction of the procedure reflected in Table 1, that the shifting refers to the iteration of the process itself, it most probably should be understood as explaining both the iteration of the process and the iteration of the shifting.

If the ambiguity and ellipticity make the verse difficult to read, one should not neglect the simplicity of the algorithm stated in such a way. Its core is pointed out; it is a succession of squarings and doublings.

This is how, step by step, we reconstruct the squaring process (for $a.10^2 + b.10 + c$):

Step 1 Squaring the last digit ($a^2.10^4$);

Step 2 Computing the successive products of 2 times the last digit with the remaining digits ($2ab.10^3$ and $2ac.10^2$);

Step 3 Adding the successive products, according to their respective powers of 10 to the partial square ($a^2.10^4 + 2ab.10^3 + 2ac.10^2$);

Step 4 Erasing the last digit, and "shifting". Then starting the process again, until no more digits of the initial number are left. (Reiterating the process with the number $b.10 + c$, then c, considering each time the partial square found in Step 3).

This hypothetical construction is illustrated in Table 1. Comparing it with other processes known in Sanskrit mathematical literature would have enabled us to justify the way we have presented it. For instance, as the process begins by squaring the last-term, we have inferred that this involved erasing the term that previously entered with that label into the process.

Table 1: Squaring: a heuristic presentation

Rule	Example: squaring 125	Squaring $a.10^2 + b.10 + c$
'When one has made the square of the last term'	1^2 is the square of the last digit. This is how one would have set down the number: 1 2 5 1 − − − −	$a^2.10^4$ is computed
'one should multiply twice that very last term ⟨separately⟩ by the remaining terms'	$2 \times 2 = 4$ and $2 \times 5 = 10$. When adding these numbers according to their respective powers of 10, the disposition obtained would be: 1 2 5 1 5 − − −	$(a.10^2)^2 + 2a10^2(b.10 + c)$
'Shifting again and again'	Erasing the digit which previously started the computation: 2 5 1 5 − − −	$b.10 + c$ is now the number to be squared
'when one has made the square of the last term'	$2^2 = 4$ is the square of the last term. When adding this quantity according to its power of 10 to the partial square found, the disposition obtained would be: 2 5 1 5 4 − −	$(a.10^2)^2 + 2a10^2(b.10 + c) + (b.10)^2$

'one should multiply twice that very last term ⟨separately⟩ by the remaining terms'	$(2 \times 2) \times 5 = 20$ is computed. When adding these values according to the respective decimal places, and placing them: $$\begin{array}{ccccc} & & & 2 & 5 \\ 1 & 5 & 4 & - & - \\ & & +2 & 0 & - \\ \hline 1 & 5 & 6 & 0 & - \end{array}$$	$(a.10^2)^2 + 2a10^2(b.10 + c) +$ $(b.10)^2 + 2b10.c$
'Shifting again and again'	When erasing the digit which previously started the computation: $$\begin{array}{ccccc} & & & 5 & \\ 1 & 5 & 6 & 0 & - \end{array}$$	
'when one has made the square of the last term'	$5^2 = 25$. When adding this value to the partial square found according to its power of ten, and placing it: $$\begin{array}{ccccc} & & & 5 & \\ 1 & 5 & 6 & 2 & 5 \end{array}$$	$(a.10^2)^2 + 2a10^2(b.10 + c) +$ $(b10)^2 + 2b10.c + c^2$
	The process ends here as there are no more digits. The square obtained is: 15625	$(a.10^2 + b.10 + c)^2$

A.2.2   Cubing

No extensive rule for cubing is given by Bhāskara in his commentary to the latter part of Ab.2.3. Cubing appears in the text as a natural extension of squaring. He quotes the beginning of a verse that recalls the structure of the verse he gave for the squaring of numbers:

> *atrāpi yeṣāṃ "antyapadasya ghanaṃ syāt" ityādi lakṣaṇasūtram, teṣām ekādīnāṃ ghanasaṅkhyā vaktavyā*
> In this case also, the cube-numbers of those ⟨digits⟩ beginning with 1 are to be recited ⟨by those⟩ whose rule which is a characterization is 'the cube of the last place should be, etc.'.

We can, however, infer the successive steps of the procedure involved, some of which may have seemed to the practitioners part of the natural process of computing (cubing $a.10 + b$):

Step 1 Cubing the last digit ($a^3.10^3$);

Step 2 Computing the successive products of 3 times the square of the last digit with the remaining digits ($3a^2b.10^2$); and adding the successive products, according to their respective powers of 10, to the partial cube ($a^3.10^3 + 3a^2b.10^2$).

Step 3 Computing the successive products of 3 times the last digit with the squares of the remaining digits ($3ab^2 10$); and adding the successive products, according to their respective powers of 3, to the partial cube ($a^3.10^3 + 3a^2b.10^2 + 3ab^2 10$).

Step 4 Erasing the last digit, and "shifting". Then starting the process again, until no more digits of the initial number are left. The partial cube considered being the one found in Step 3.

This hypothetical computation is illustrated in Table 2.

Table 2: Cubing 63

Hypothetical rule	The cubing of 63	The cubing of $a.10 + b$
cube the last digit	$6^3 = 216$. The disposition would be: 6 3 2 1 6 – – –	$(a.10)^3$

Table 2: Cubing 63

Considering the successive product of 3times the square of the last digit with the remaining digits	As $3 \times 6^2 \times 3 = 324$, the disposition would be:
	
	           6 3
	2 1 6 — — —
	   3 2 4 — —
	
	Adding according to the respective decimal places of each digit:
	
	           6 3
	2 4 8 4 — —
		$(a.10)^3 + 3a^2 10^2 .b$

Computing successively the product of 3 times the last digit with the square of the following digits	As $3 \times 6 \times 3^2 = 162$, the disposition would be:
	
	           6 3
	2 4 8 4 — —
	       1 6 2 —
	
	Adding according to the respective decimal places of each digit:
	
	           6 3
	2 5 0 0 2 —
		$(a.10)^3 + 3a^2 10^2 .b + 3a10.b^2$

Erasing the last digit	3 2 5 0 0 2 —	Considering that the number to cube is b.

Table 2: Cubing 63

| Cubing the next digit | As $3^3 = 27$, the disposition would be:

$$\begin{array}{cccccc} & & & & & 3 \\ 2 & 5 & 0 & 0 & 2 & - \\ & & & & 2 & 7 \end{array}$$

Adding according to the respective places of each digit:

$$\begin{array}{cccccc} & & & & & 3 \\ 2 & 5 & 0 & 0 & 4 & 7 \end{array}$$ | $a^3.10^3 \quad + \quad 3a^2 10^2.b \quad +$ $3a10.b^2 + b^3$ |
| As there are no more digits the process ends here. The cube found is therefore: | 250047 | $(a.10 + b)^3$ |

A.3 Squaring and cubing with fractions

The number $a + \frac{b}{c}$ is noted in this edition of Bhāskara's commentary in the following way[10]: $\begin{array}{c} a \\ b \\ c \end{array}$.

This is what is called in this part of the text 'a fraction' (*bhinna*).

The computation of the square of fractions is described here in two sequences. Firstly:

> *bhinnavargo 'py evam eva| kintu sadṛśīkṛtayoś chedāṃśarāśyoḥ pṛthak pṛthag vargaṃ kṛtvā chedarāśivargeṇāṃśarāśivargasya bhāgalabdhaṃ bhinnavargaḥ|*
>
> The square of fractions is also just like this. However, when one has made separately the squares of the numerator and denominator quantities, that were made into the same kind, the result of the division of the square of the numerator quantity by the denominator quantity is the square of the fraction.

---

[10]One should keep in my mind that this is the way manuscripts note fractions. Moreover, the notations adopted in manuscripts may have been different from those used by Bhāskara, more than 1000 years earlier.

Secondly:

> *chedaguṇaṃ sāṃśam iti*|
> '⟨the whole number⟩ having the denominator for multiplior increased
> by the numerator'

Probably the expression used in the first sequence: '...(the numerator and denominator) are made into a same kind', refers to the operation described in the second sequence. This operation transforms the fractionary number given in the problem into a fraction with just a denominator and a numerator. Indeed, if we consider simultaneously the general notations we have adopted and the quantity $(6 + \frac{1}{2})$ treated in detail by Bhāskara in Example 2, we can infer the following computation[11]:

$$\begin{array}{c|c} a & 6 \\ b & 1 \\ c & 4 \end{array}$$

becomes ('⟨the whole number⟩ having the denominator for multiplior')

$$\begin{array}{c|c} ac & (6 \times 4) \\ b & 1 \\ c & 4 \end{array}$$

'increased by the numerator'

$$\begin{array}{c|c} ac + b & (6 \times 4) + 1 \\ c & 4 \end{array}$$

If we follow then, the first sequence for squaring fractions: $\dfrac{(ac+b)^2}{c^2} \left| \dfrac{25^2}{4^2} \right.$

The numerator is then divided by the denominator:

$$(ac + b)^2 = q.c^2 + r \mid 625 = 39 \times 16 + 1$$

The result obtained is noted as a fractionary number: $\begin{array}{c|c} q & 39 \\ r & 1 \\ c^2 & 16 \end{array}$.

No rule is given concerning a whole number decreased by a part, however such a fraction appears in Example 2.

The cubing of fractions is, as is the case for the cubing of whole numbers, referred to briefly as a mere extension of the process for squaring fractions. Please see Table 3 for how we guess this was carried out.

---

[11]We do not know how the intermediary steps were presented, this whole presentation is therefore arbitrary.

Table 3: Cubing a fraction

Example 4 of BAB.2.3.cd is stated as follows:

*ṣaṭpañcadaśāṣṭānāṃ   tāvadbhāgair   vihīnagaṇitānām| ghanasaṅkhyāṃ   vada
viśadaṃ yadi ghanagaṇite matir viśadā||*
4. Say, clearly, the cube-number of six, five, ten and eight that are computed
as decreased by their respective parts|
If ⟨you have⟩ a clear knowledge in cube-computations||

The fractions considered in the text are, for us, of the following form:

$$6 - \frac{1}{6} = 5 + \frac{5}{6}$$

$$5 - \frac{1}{5} = 4 + \frac{4}{5}$$

$$10 - \frac{1}{10} = 9 + \frac{9}{10}$$

$$8 - \frac{1}{8} = 7 + \frac{7}{8}$$

This set of numbers, which is equal in value to the one above, is set down:

```
5    4    9    7
5    4    9    7
6    5    10   8
```

Let us consider the process involved in the cubing of the last fraction of this

example:
$$\begin{array}{c} 7 \\ 7 \\ 8 \end{array}$$

This column of numbers, representing the number $7 + \frac{7}{8}$, should be first transformed
into a form with numerator and denominator only.

That is into $\dfrac{63}{8}$.

Then, the cubes of the numerator and denominator are made separately.
The hypothetical steps followed for cubing 63 (the result found is 250047) are
illustrated in Table 2.The cube of 8 (512) is given in the resolution of Example
3 of BAB.2.3.cd.

Table 3: Cubing a fraction

Dividing the cube of the numerator by the cube of the denominator:
$\frac{250047}{512} = 488 + \frac{191}{512}$, which corresponds to the last column set down as a result: 488 191 . 512

B   BAB.2.4-5

B.1   Extracting square-roots

B.1.1   Square and non-square places

The procedure of square root extraction rests upon a categorization of the places of the decimal place-value system (defined in AB.2.2). Āryabhaṭa distinguishes square (*varga*) and non-square (*a-varga*) places. A square place is one which stands for an even power of ten (e.g. $10^0, 10^2, 10^4,\ldots$). A non-square place stands for a power of ten which is not a square (e.g. $10^1, 10^3, 10^5,\ldots$).

Bhāskara substitutes for it his own categorization. He considers the places where the digits forming the number whose root is to be extracted are to be noted. He counts them from right to left, distinguishing between places associated to an even number and places associated to an odd number. The place for the digit whose power of ten is 10^0 is the first to be counted, therefore the so-called "square" places are found for all odd numbers of places, and the so-called "non-square" places for all even numbers of places. This is for instance how both categorize the places associated to 625 (whose square-root is extracted in Example 1 of BAB.2.4 and whose extraction is illustrated in Table 4[12]:

$odd(3)$	$pair(2)$	$odd(1)$
10^2	10^1	10^0
v	av	v
6	2	5

B.1.2   The procedure

The detail of the procedure and how precisely it was carried out is not known to us. A heuristic reconstruction is given in Table 4. In the following, we will consider

---

[12]For a brief analysis of the way the rule is composed see the Introduction in Volume I section 2.3.

that the digits forming a number are ordered from left to right: the first digit
being the one standing in the highest place. The steps that we have reconstructed
– some of which may have seemed so natural that it wasn't deemed necessary to
state them – may be summed up as follows:

Step 1 Probably by trial and error, find the biggest square (a^2) smaller than the
first digit. (Or the biggest square smaller than a two digit number, if the last
digit does not fall on a place standing for a square power of ten).

Step 2 Subtract it from the last digit, and substitute the difference in place of
the former digit. The square-root of this square (a) is the last digit of the
square-root sought.

Step 3 Considering the next place to the right, divide the number formed by
considering all the digits to the left of that place (that place included) by
twice the partial square-root obtained.

Step 4 Replace the dividend by the remainder of the division. The quotient is
considered here to be the next digit of the square-root sought. In fact it is
either the quotient or the quotient increased by 1, which is the next digit of
the square-root to be extracted. Bhāskara never goes into the detail of his
root extractions, therefore we do not know if he was aware of such a step.

Step 5 Considering the next place to the right, subtract from the number formed
by all the digits to the left of that place (that place included) the square of
the quotient. Replace that number by the difference. Re-iterate the process
starting from Step 3. The process ends when one cannot shift to the right
anymore.

Among the steps that are neither mentioned by Āryabhaṭa nor by Bhāskara, we
can list:

- The way the square-root of the first digit (or two-digit number when the last
digit of the number whose root is to be extracted falls in a non-square place)
is found is not mentioned.

  We can note here that both Āryabhaṭa and Bhāskara, by not indicating how
  the procedure starts, seem to emphasis its iterative quality.

- The place where the successive digits of the square-root extracted are placed
is not mentioned. Later authors have indicated that they should be noted on
a separate line. Bhāskara may be referring to such a line when he comments
on the compound *stānāntare* (in a different place) used in Ab.2.4:

  > *sthānād anyasthānaṃ sthānāntaram, tasmin sthānāntare tasya lab-*
  > *dhasya mūlasaṃjñā| yatra punaḥ sthānāntaram eva na vidyate,*
  > *tatra tasya tatraiva mūlasaṃjñā|*
  > A place other than the ⟨given⟩ place is a different place ; in this

different place, the quotient has the name root. When, however,
a different place precisely does not exist, then that ⟨quotient⟩ has
the name root in that very place ⟨where it was obtained⟩.

There are two ways of understanding this sentence: it may refer to the shifting
to a different place in the decimal place value notation used to set down the
digits. It may also here indicate a separate space on the working surface where
the digits of the square-root extracted appear progressively as the process
follows its way. The sybillin last sentence of this paragraph, in both cases,
refers to the way the process ends. If it concerns the space where the digits of
the square-root extracted appear, it may mean that in the case of a square-
root found at once (as for digits or two digit numbers) no separate space is
needed. Among the other steps not specified by Āryabhaṭa or Bhāskara we
can note:

- When the division is performed, the remainder replaces the digits that formerly entered the division as dividend. This may have been a regular feature
of the division procedure[13].

- The way that the intermediary operations of placing the remainder, the result
of the subtraction etc, are noted and how they interplay with their respective
powers of ten is not indicated either. This may also have been a feature of
computation considered as self-evident.

Table 4: Extracting the square-root of a three digit number

Āryabhaṭa's rule	Example: extracting the square root of 625	Extracting the square-root of $A = (a.10 + b)^2$
When subtracting the square from the square ⟨place⟩	The biggest square smaller than 6, which is the digit in the "highest square place", is 4. So that 2 is the first digit of the square-root to be extracted. This is how the number may have been set down: $\quad v \quad av \quad v$ $\quad 6 \quad 2 \quad 5$ $-4 \quad - \quad -$ $\quad 2 \quad 2 \quad 5$	$A - a^2.10^2$ is computed. $a.10$ is the partial square-root extracted.

---

[13] See for instance [Datta&Singh 1938; p.152]

Table 4: Extracting the square-root of a three digit number

Āryabhaṭa's rule	Example: extracting the square root of 625	Extracting the square-root of $A = (a.10 + b)^2$
One should divide, constantly, the non-square ⟨place⟩ by twice the square-root.	22 is considered to be in the 'non-square' place. Twice the partial square root is $2 \times 2 = 4$. One performs the following division: $$\frac{22}{4} = 5 + \frac{2}{4}.$$ 5 is the quotient, it is the second digit of the square-root to be extracted. The partial square-root is, at this point: 25. The remainder of the division of 22 by 5 is set down in the place of the previously written digits: $$\begin{array}{cc} av & v \\ 2 & 5 \end{array}$$	b is computed as the quotient of the division of the two higher digits by a^2. Then $A - a^2.10^2 - 2ab10$ is set down. $a.10 + b$ is the partial square-root extracted.
The quotient is the root in the next place. When subtracting the square from the square	The quotient is 5. The next place being a square-place, one subtracts the square of 5. $$\begin{array}{cc} av & v \\ 2 & 5 \\ & -5^2 \\ & 0 \end{array}$$	$A - a^2.10^2 - 2ab10 - b^2$ is computed.
The square-root found is	25	$a.10 + b$

B.2 Extracting cube-roots

B.2.1 Cube and non-cube places

As for square-root extraction, the cube root extraction procedure uses a categorization of the places of the decimal place-value system: there are cube (*ghana*) and non-cube (*aghana*) places. They form an ordered set. Āryabhaṭa's rule refers to a first and a second non-cube place. In BAB.2.5., Bhāskara glosses as well:

> *atra gaṇite ghana ekaḥ, dvāvaghanau|*
> In this computation, there is one cube, two non-cube ⟨places⟩.

These names correspond to the respective power of tens of the places: a cube place is a place whose power of ten is a multiple of three (e.g. $10^0, 10^3, 10^6, \ldots$); a non-cube place is a place whose power of ten is not a multiple of three (e.g. $10^1, 10^2, 10^4, \ldots$). The place for 10^0 is considered to be a cube place. The second non-cube place is the second from the right in the sub-triplet of the ordered set made of (a cube place, a non-cube place, a non-cube place).

This categorization is illustrated with the number 1728 (whose cube-root is extracted in Example 1 of BAB.2.5 and whose extraction is shown in Table 5):

$$
\begin{array}{cccc}
10^3 & 10^2 & 10^1 & 10^0 \\
g & a-g & a-g & g \\
1 & 7 & 2 & 8
\end{array}
$$

B.2.2 The procedure

We do not know precisely how each step of the procedure was carried out. We have presented heuristically a reconstruction of the procedure in Table 5, although this would need to be justified and be compared with other procedures known to us from the Sanskrit tradition. In this reconstruction, the digits of the number whose cube-root is considered are considered from left to right. The first digit is therefore the one which stands for the multiple of the highest power of ten.

Step 1 Find, probably by trial and error, the biggest cube smaller than the first digit. (Or smaller than a two-digit/three digit-number if the first digit of the number whose root is extracted does not fall on a place whose power of ten is a cube.)

Step 2 Subtract the cube from the first digit (or from the two-digit/three-digit number). Replace the digit (resp. two-digit/three-digit number) by the difference. The cube-root of the subtractor is the first digit of the cube-root sought.

Step 3 Shift by one place to the right. Compute the product of three times the square of the partial cube-root obtained. Divide the number obtained by considering all the digits to the left of this place (this place included) with the previous product. Erase the number and replace it with the remainder of the division. The quotient is considered to be the next digit of the cube-root sought. In fact once again, this may not be exactly the right digit and one may have to increase by one or by two so that the computation remains correct. However we do not know if this step was carried out in such a way.

Step 4 Considering the next place to the right, compute the product of three times
the square of the quotient with the partial cube-root obtained before Step 3.
Subtract from the number obtained by considering all the digits to the left
of that place (that place included) the product. Replace that number with
the difference obtained.

Step 5 Shift by one place to the right. Subtract from the number obtained by
considering all the digits to the left of this place (this place included) the
cube of the quotient obtained in Step 3. Reiterate the process starting with
Step 3. The process ends when one cannot shift to the right anymore.

Among the steps that are neither mentioned by Āryabhaṭa nor by Bhāskara, we
can list:

- The way the cube-root of the first digit (or two/three-digit number when the
  last digit of the number whose root is to be extracted falls in a non-cube
  place) is found is not given by either of the two authors. This step involves
  finding the greatest cube smaller than that digit (or two/three-digit number).

  Once again, this may be a way of emphasizing the iterative quality of the
  procedure.

- The space where the successive digits of the cube-root extracted are placed is
  not referred to. Later authors have indicated that they should be noted on a
  separate line. If this was suggested elliptically in the commentary to Ab.2.4.,
  it may then be assumed here.

- We include in this list an elliptic formulation:

  > The square of ⟨the quotient⟩ multiplied by three and the former
  > ⟨quantity⟩ should be subtracted from the first ⟨non-cube place⟩

  Though nowhere explained the "former ⟨quantity⟩" is the partial cube-root
  obtained, before the computation of the quotient (the quotient obtained before Step 3 in our presentation).

- The fact is that when the division is performed, the remainder replaces the
  digits that formerly entered the division as dividend. As in the process described in BAB.2.4., this may be a regular feature of the division procedure.

- The way that the intermediary operations of placing the remainder, the result
  of the subtraction etc, are noted and how they interplay with their respective
  powers of ten is not indicated. This may also be a feature of the computation,
  considered as so usual that it was not thought to have to be described.

Table 5: An example of the procedure for extracting the cube-root

Āryabhaṭa's rule	Example: extracting the cube-root of 1728	Extracting the cube-root of $A = (a.10 + b)^3$
'And the cube ⟨should be subtracted⟩ from the cube place'	The digit in the "cube place" is 1, the highest cube smaller than 1 is 1^3, it is subtracted in the cube placed, and replaced by the result: $$\begin{array}{cccc} g & a-g & a-g & g \\ 1 & 7 & 2 & 8 \\ -1 & - & - & - \\ 0 & 7 & 2 & 8 \end{array}$$ 1 is the first digit of the partial cube-root extracted.	$A - (a.10)^3$ is computed. $a.10$ is the first digit of the cube-root extracted.
One should divide the second non-cube place by three times the square of the root of the cube	The digit in the second non-cube place is 7. The square of the root of the former cube is 1^2 $$\frac{7}{3} = 2 \times 3 + 1$$ The remainder of the division of 7 by 3 replaces the digit of the "second non-cube place": $$\begin{array}{cccc} g & a-g & a-g & g \\ & 7 & 2 & 8 \\ & 1 & 2 & 8 \end{array}$$ 12 is the partial cube-root extracted.	b is found as the quotient of the division of the digit of the second non-cube place by $3a^2$. $A - [(a.10)^3 - 3a^2 10^2.b)]$ is computed. $a.10 + b$ is the partial cube-root extracted.

Table 5: An example of the procedure for extracting the cube-root

Āryabhaṭa's rule	Example: extracting the cube-root of 1728	Extracting the cube-root of $A = (a.10 + b)^3$
The square of ⟨the quotient⟩ multiplied by 3 and the former ⟨quantity⟩ should be subtracted from the first ⟨non-cube place⟩	The "former quantity" considered here is the first digit of the cube-root found: 1, the quotient of the division of 7 by 3 is $2 \times 3 \times 2^2 \times 1 = 12$. This is subtracted: $\begin{array}{cccc} g & a-g & a-g & g \\ 1 & & 2 & 8 \\ -1 & & 2 & - \\ - & & - & 8 \end{array}$	$A - [(a.10)^3 - 3a^2 10^2.b - 3a.10.b^2)]$ is computed.
and the cube from the cube ⟨place⟩	The digit in the cube-place is 8. The cube of the quotient is 2^3. $\begin{array}{cccc} g & a-g & a-g & g \\ & & & 8 \\ & & & -2^3 \\ & & & 0 \end{array}$ The process ends here as there are no more digits. The cube-root extracted is 12.	$A - [(a.10)^3 - 3a^2 10^2.b - 3a.10.b^2) - b^3]$ is computed. The cube-root extracted is $a.10 + b$

C BAB.2.6

C.1 Area of a triangle

Āryabhaṭa's rule, according to Bhāskara's interpretation[14] concerns a general case:

> Ab.2.6.ab. The bulk of the area of a trilateral is the product of half the
> base and the perpendicular|

This can be understood as follows:

---

[14]Because Āryabhaṭa uses the compound *samadalakoṭī* or "halving upright", probably this rule was intended originally only for equilaterals and isosceles.

Figure 7: Equilateral and isoceles triangles

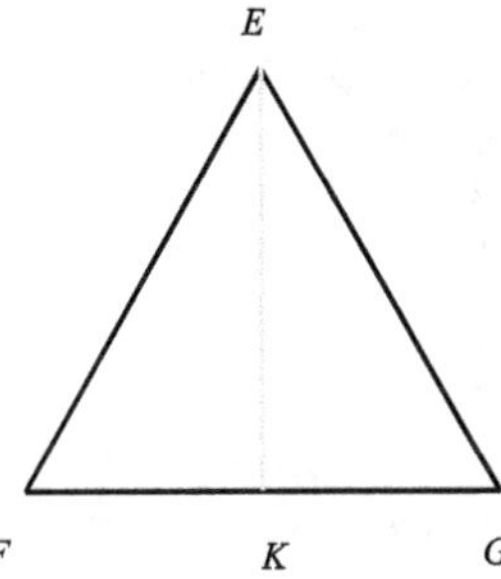

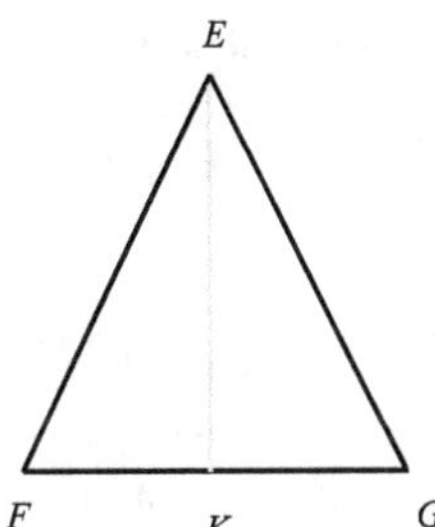

As illustrated in Figure 8, let MNO be a triangle. If MD is the height issued from M and falling on the base NO, then the area $\mathcal{A}$ of MNO will be

$$\mathcal{A} = \frac{NO}{2} \times MD.$$

C.1.1 Equilaterals and isosceles triangles

Bhāskara gives in his commentary to Example 1 of Ab.2.6.ab a property of equilateral triangles:

> *samatryaśrikṣetre samaivāvalambakasthitiḥ iti*
> 'In an equi⟨lateral⟩ trilateral field the location of the perpendicular is precisely equal.'

In other words, in an equilateral triangle any height sections the corresponding base into two equal segments.

This is also stated for isosceles triangles:

> *dvisamatryaśrikṣetrasyāpi 'samaivāvalambakasthitiḥ' iti*
> For an isosceles trilateral also, 'the location of the perpendicular is precisely equal.'

This property is used along with Ab.2.17 which states the so-called 'Pythagoras Theorem' to justify the following procedure:

Problem Knowing the length of the sides of an equilateral or isosceles triangle, find its area. Let EFG be such an equilateral triangle illustrated in Figure 7, which also shows an isoceles triangle.

Step 1 Compute the length of any height in the case of an equilateral, issued from the vertex in the case of an isosceles triangle.

Using the property stated above, if EK is the perpendicular issued from D onto FG, we know that

$$FK = KG = \frac{EF}{2}.$$

Bhāskara specifies when quoting Ab.2.17:

> *'yaś ca eva bhujāvargaḥ koṭīvargaś ca karṇavargaḥ sa [Ab.2.17] iti bujākoṭyor vargau karṇavargaḥ| tena bhujāvarge karṇavargāc chuddhe śeṣaṃ samadalakoṭīvargaḥ*

> 'That which is the square of the base and the square of the height is the square of the hypotenuse'.
> Therefore, the square of the hypotenuse is ⟨produced with⟩ the squares of both the base and the height. Hence, when the square of the base is subtracted from the square of the hypotenuse, the remainder is the square of the perpendicular. . ..

In other words

$$EK^2 = EF^2 - FK^2 = EG^2 - KG^2.$$

Step 2 In the cases observed, the square found for the length of the perpendicular is not perfect – i.e, its square root cannot be extracted without an approximation. Therefore, the length of half the base is squared so that it can enter the rule given by Āryabhaṭa. In other words $\frac{FG^2}{4}$ is computed.

Step 3 The rule given in the verse is applied:

$$\mathcal{A}^2 = \frac{EK^2 \times FG^2}{4} \Leftrightarrow \sqrt{\mathcal{A}^2} = \sqrt{\frac{EK^2 \times FG^2}{4}}.$$

The square-root expression is written here to recall the double meaning that the word *karaṇī* may take here.

C.1.2 Uneven triangles

Bhāskara uses the following property of the lengths of any triangle:

> *bhujayor vargaviśeṣaḥ tayor vā samāsaviśeṣābhyāsaḥ*
> *tribhujakṣetre ābādhāntarasamāsaviśeṣābhyāsbhavati*
> In a trilateral field the difference of the squares of the two sides, or the product of the sum and the difference of the two, is the product of the sum and the difference of ⟨its⟩ different sections of the base.

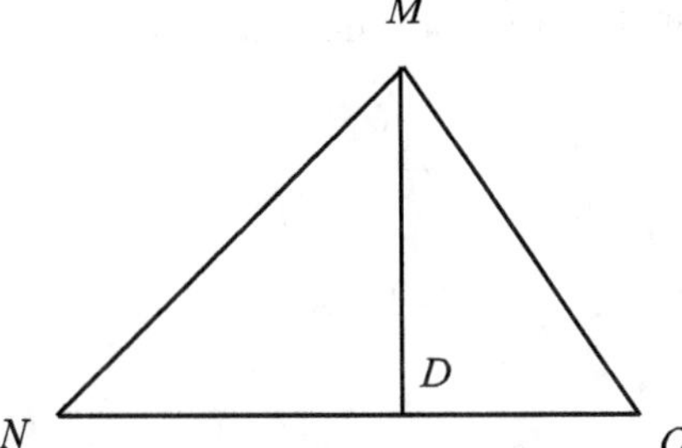

Figure 8: Any triangle

In other words:

Let MNO be any triangle such as is illustrated in Figure 8, let MD be a height.

The sections of the base are the two segments ND and DO for the sides MN and MO. The first sentence of this paragraph may be translated in our algebraical language as

$$MN^2 - MO^2 = (MN + MO)(MN - MO) = (ND + DO)(ND - DO)$$

The last equality, which may also be stated as $MN^2 - MO^2 = ND^2 - DO^2$, is easily derived from the "Pythagoras Theorem".

Bhāskarra then writes:

> *bhūmyā ābādhāntarasamāsapramāṇayā vibhajya labdhaṃ bhūmāv eva saṃkramaṇam|*
> When one has divided by the base whose size is the sum of ⟨its⟩ different sections, a *saṃkramaṇa* is ⟨applied⟩ to the same base together with the quotient.

Dividing the above equalities by the base:

$$\frac{MN^2 - MO^2}{NO} = \frac{(ND + DO)(ND - DO)}{NO}.$$

Since the base is the sum of its segments, $NO = ND + DO$, then

$$\frac{MN^2 - MO^2}{NO} = ND - DO.$$

In the *saṃkramaṇa* operation, stated in Ab.2.24, this quantity is considered under the name "quotient" (*labdha*):

$$x = \frac{MN^2 - MO^2}{NO} = ND - DO.$$

It is used along with the size of the base, $NO = ND + DO$, which is the quantity that is 'increased or decreased'. The *saṃkramaṇa* operation can be understood, then, as the computation of the two following quantities:

$$u = \frac{NO + x}{2}$$

and

$$v = \frac{NO - x}{2}.$$

One can easily check that

$$u = ND$$

and

$$v = DO.$$

With either one of these different segments of the base, it is understood that one can follow the method described above for equilateral and isosceles triangles to reckon the perpendicular's length, and from there compute the area of the triangle.

The different steps of the procedure to be followed are therefore:

Problem Knowing the lengths of the sides of triangle MNO, find the area.

Step 1 Compute

$$x = \frac{MN^2 - MO^2}{NO}.$$

Step 2 Use a *saṃkramaṇa* in order to find the lengths of the two different sections of the base:

$$\frac{NO + x}{2} = ND,$$

$$\frac{NO - x}{2} = DO.$$

Step 3 Find the length of the perpendicular by either one of the following computations:

$$AD^2 = MN^2 - MD^2 = NO^2 - DO^2.$$

Step 4 The area is

$$\mathcal{A}^2 = \frac{MD^2 \times NO^2}{4} \Leftrightarrow \sqrt{\mathcal{A}^2} = \sqrt{\frac{MD^2 \times NO^2}{4}}.$$

Figure 9: An equilateral pyramid with a triangular base

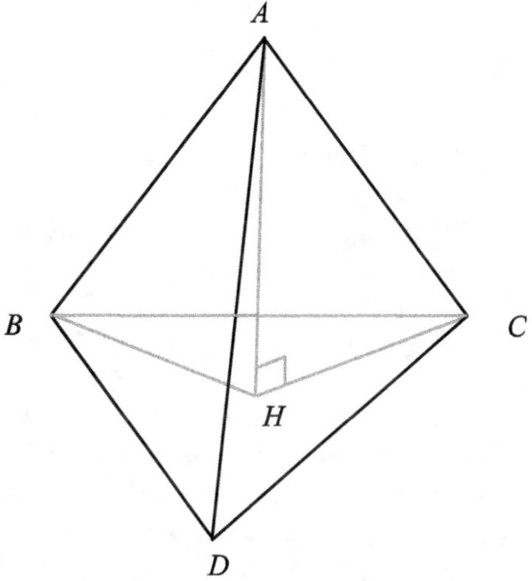

C.2 Volume of a pyramid

C.2.1 General rule

The rule given by Āryabhaṭa, in the second half of verse 6, is interpreted by Bhāskara as giving the volume of a triangular based equilateral pyramid. We may relate the relation given here as follows:

Given an $ABCD$ pyramid, illustrated in Figure 9, AH is the perpendicular issued from A onto the triangle BDC. If the area of BDC is $\mathcal{A}$, then the volume $\mathcal{V}$ of $ABDC$ is

$$\mathcal{V} = \frac{1}{2}\mathcal{A} \times AH.$$

This formula for the volume of a pyramid is incorrect.

The correct formula is

$$\mathcal{V} = \frac{1}{3}\mathcal{A} \times AH.$$

Although we do not know why and how Āryabhaṭa derived this wrong relation, we can make the following hypothesis: the solid equilateral is probably seen as deriving geometrically from the area by the same process that derives from two lines a surface. This continuity between the two-dimensional field and the three-dimensional field may be the key to the relation given here. As Ab.2.6ab. derives the area of an equilateral triangle by the product of half the base and the height, the volume of the pyramid seems to be derived by half its base (which is here the area of an equilateral triangle) and its height.

C.2.2   A śṛṅgāṭaka

Bhāskara gives the following description and explanation[15]:

> *ūrdhvabhujā hi nāma kṣetramadhye ucchrāya iti pratyakṣam| sa ca tirya-
> gavasthitasya śṛṅgāṭakakṣetrabāhoḥ karṇavadavasthitasya koṭiḥ| bhujā
> karṇamūlakṣetrakendrāntāralam*

> It is obvious (*pratyakṣa*) that the so-called "upward-side" is a height in
> the middle of the field. And that is the upright-side (*koṭi*) for the side of
> a *śṛṅgāṭaka* field which is located obliquely as an ear, ⟨while⟩ the base
> is the intermediate space in between the root of the ear and the center
> of the field.

Let $ABCD$ be an equilateral triangular based pyramid as represented in Figure
10. Let AH be the height issued from A and falling onto the triangle BCD.

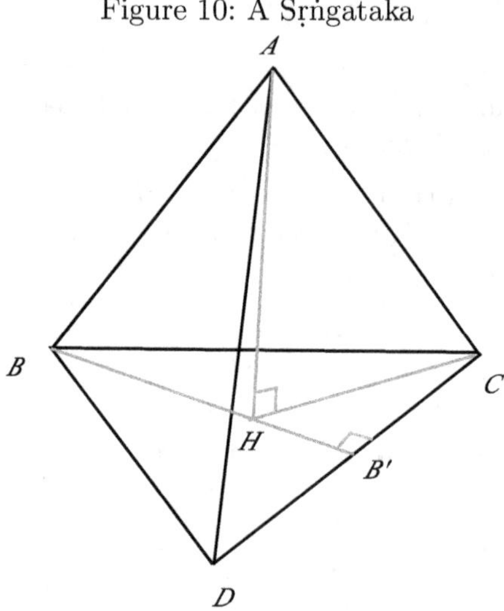

Figure 10: A Śṛṅgataka

AC is what is called the ear (*karṇa*), it is also the hypotenuse of AHC. AH is
what is called the upward-side (*ūrdhvabhujā*) of the *śṛṅgāṭaka*. It is defined at the
beginning of the commentary of this half-verse:

---

[15][Shukla 1976; p.48, lines 8-10]

ūrdhvabhujā kṣetramadhye ucchrāya
The upward side is a height in the middle of the field.

And this sentence is recalled at the beginning of the text quoted above.

sa ca tiryagavasthitasya śṛṅgāṭakakṣetrabāhoḥ karṇavadavasthitasya koṭiḥ[16]
and that is the upright-side for the side of a *śṛṅgāṭaka* field which is located obliquely as an ear....

CH, in the above quoted text, is called the base (*bhujā*).

In the resolution of Example 1, Bhāskara writes:

labdho 'ntaḥkarṇaḥ [karaṇyaḥ] 48| ayam eva karṇaḥ
ūrdhvam avasthitatribhuja[kṣetrasya bhujā]|
The inner ear obtained is 48 [*karaṇīs*]. This very ear is the [base] of the
trilateral [field] located upwards.

So that here CH is referred to both as an inner-ear (*antaḥkarṇa*) – that is as the
hypotenuse of $CB'H$ – and as the base of the right-angle triangle AHC. The word
base has been added in brackets by the editor as all manuscripts, except one, omit
this word.

The first text quoted in this section is the part of the commentary where the word
śṛṅgāṭaka appears for the first time. Because it is used in examples to refer to
the pyramid itself, we understand it as the name of an equilateral pyramid with
a triangular base *with* a perpendicular issued from one top to the center of the
triangular base.

C.2.3 A Rule of Three

The computation of CH, from which the upright side AH may be computed, rests
upon the proportional properties of similar triangles.

Bhāskara states such properties by formulating them through a Rule of Three:

tadānayane trairāśikam- yadi tribhujakṣetrāvalambakena
tribhujakṣetrabāhur labhyate tadā tasyaiva
tribhujakṣetrabāhudalasaṅkhyakasyāvalambakasya
kiyān bāhur iti
When computing that ⟨base⟩, a Rule of Three: 'If the side of a trilateral
field is obtained with the perpendicular of that very trilateral field,
then for the perpendicular whose amount is half the side of the ⟨initial⟩
trilateral field, how much is the side?'

This can be understood as follows, as illustrated in Fig 11, next page.

---

[16] Two manuscrits read *koṭī*. However it is also the upright-side (*koṭi*) of the right-angle triangle
(AHC), which would be in accordance with the regular use of the word.

Figure 11: Rule of Three

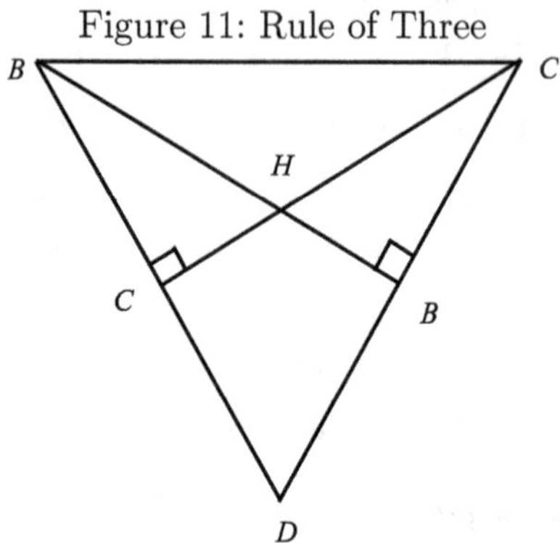

The triangles $BB'C$ and $B'CH$ are similar:

$$BB' : CB = CB' : CH.$$

So that in other words

$$CH = \frac{CB \times CB'}{BB'}.$$

Because BDC is an equilateral field, $CB' = \frac{CB}{2}$. CB' is thus 'the perpendicular whose value is half the the side of ⟨initial⟩ trilateral'.

If the lengths considered are *karaṇīs*, the square of such an equality is considered.

C.2.4   The procedure followed

Problem Knowing the side of an equilateral triangular based pyramid $ABCD$, find its volume.

Step 1 If AH is the perpendicular issued from A onto BCD, then with a Rule of Three we know that

$$CH = \frac{CB \times CB'}{BB'}.$$

If CB is a *karaṇī*, in which case BB' may be one, the following computation is in fact carried out:

$$CH^2 = \frac{CB^2 \times CB'^2}{BB'^2}.$$

Step 2 Then we use Ab.2.17ab, from which we know that:

> *karṇakṛteḥ bhujāvargaviśeṣaḥ ūrdhvabhujāvargaḥ*
> The difference of the square of the base and the square of the
> hypotenuse is the square of the upright side.

So that

$$AH^2 = AC^2 - CH^2.$$

Step 3 Then according to the rule given by Āryabhaṭa here, as Bhāskara specifies:

> *ardhamityatra karaṇitvād dvayoḥ karaṇībhiścaturbhirbhāgo hriyate|*
> Since ⟨the rule uses the expression⟩ "half", because two are *karaṇīs*,
> one should divide by four *karaṇīs*.

$$\mathcal{V}^2 = \frac{1}{4}\mathcal{A}^2 \times AH^2 \Leftrightarrow \sqrt{\mathcal{V}^2} = \sqrt{\frac{1}{2}\mathcal{A}^2 \times AH^2}.$$

D   BAB.2.7

D.1   Area of a circle

D.1.1   The general rule

Āryabhaṭa gives the following rule:

> *samapariṇāhasyārdhaṃ viṣkambhārdhahatam eva vṛttaphalam|*
> Ab.2.7.ab. Half of the even circumference multiplied by the semi-diameter,
> only, is the area of the circle|

In other words, for a circle of circumference $\mathcal{C}$ and diameter D, the area $\mathcal{A}$ is
according to this definition:

$$\mathcal{A} = \frac{\mathcal{C}}{2} \times \frac{D}{2}.$$

D.1.2   Procedure used in examples

Problem Knowing the diameter D of a circle, find its area $\mathcal{A}$.

Step 1 Using the values given in Ab.2.10, and a Rule of Three, find the (approximate) circumference $\mathcal{C}$ of the circle.

Ab.2.10 states that a circle of diameter 20 000 has a circumference of 62832.
Bhāskara indicates:

*...trairāśikena vakṣyamāṇaviṣkambhaparidhi
pramāṇaphalābhyāṃ...*

...by means of a Rule of Three, with as measure and fruit ⟨quantities⟩,
the diameter and the circumference to be told [in Ab.2.10]...

setting down a Rule of Three as described in BAB.2.26:

The measure quantity	The fruit quantity	The desire quantity
20 000	62832	D

Then the fruit of the desire, the circumference ($\mathcal{C}$) is

$$\mathcal{C} = \frac{D \times 62832}{20000}.$$

The result obtained, if it is not integer, is in the form of an integer with a
fractional part (see the procedure described in the Annex on BAB.2.3).

Step 2 Having thus the diameter and the circumference, one can then compute
the area according to Āryabhaṭa's rule:

$$\mathcal{A} = \frac{\mathcal{C}}{2} \times \frac{D}{2}.$$

The result obtained is an approximation, as Āryabhaṭa states that the ratio given
in verse 10 is one. Bhāskara does not stress this point here, on the contrary, he
insists, rightly, that the procedure, given in all its generality, is accurate.

D.2   Volume of a sphere

D.2.1   General rule

Āryabhaṭa gives the following rule:

tannijamūlena hataṃ ghanagolaphalaṃ niravaśeṣam‖

Ab.2.7.cd. That multiplied by its own root is the volume of the circular
solid without remainder.

In other words, for a sphere whose volume is $\mathcal{V}$, whose diametral subsection has
an area $\mathcal{A}$, the volume would be

$$\mathcal{V} = \mathcal{A} \times \sqrt{\mathcal{A}}.$$

Bhāskara reinterprets the rule as follows, because in most cases the square-root of
the area cannot be obtained exactly:

> *tat punaḥ kṣetraphalaṃ mūlakriyamāṇaṃ*
> *karaṇitvaṃ pratipadyate yasmāt karaṇīnāṃ mūla[mapekṣitam]|*
> *tataḥpunar api karaṇīnām akaraṇībhiḥ saṃvargo nāstīti*
> *kṣetraphalaṃ karaṇyate| evam ayam artho 'rthād avasīyate*
> *kṣetraphalavargaḥ kṣetraphalena guṇita iti|*

On the other hand, that area becomes a *karaṇī* when being made into a root (*mūlakriyamāna*), because a root is [required] of a square (*karaṇī*). However, also, as there is no product of a *karaṇī* by a non-*karaṇī*, the area of the field is made into a *karaṇī*/the area of the field is squared (*karaṇyate*). Consequently, the following meaning is understood in fact: the square (*varga*) of the area of the field is multiplied by the area of the field.

Following Bhāskara's interpretation, with the same notation as before, this is the computation to be used:

$$\mathcal{V}^2 = \mathcal{A}^2 \times \mathcal{A}.$$

Bhāskara discusses another rule, dismissed as "practical" (*vyāvahārika*):

> *vyāsārdhaghanaṃ bhittvā*
> *navaguṇitam ayoguḍasya ghanagaṇitam|*

When one has halved the cube of half the diameter and multiplied by nine, the computation of the volume (*ghanagaṇita*) of the sphere (*ayoguḍa* lit. iron ball) ⟨is obtained⟩ |

In other words, for a sphere whose volume is $\mathcal{V}$, whose diameter is D, the 'practical' volume would be

$$V = \frac{9 \times (\frac{D}{2})^3}{2}.$$

This relation given by Āryabhaṭa for the volume of a sphere is incorrect, as well as the one quoted by Bhāskara. The correct rule is, if the diameter of the sphere is $D = 2R$:

$$V = \frac{4}{3}\pi R^3 = \frac{2}{3} \times D \times \mathcal{A}.$$

We do not know how this rule was derived, nor why this specific wrong relation was considered. As in the case of an equilateral triangular based pyramid, it may have been linked to the conception of the geometrical derivation of the solid from the surface: that of the "product" of a height on the disk.

D.3 Procedure followed in examples

Problem Knowing the diameter D of a sphere, compute its volume.

Step 1 Compute the area, $\mathcal{A}$ of the diametral section, according to the procedure described above.

Step 2 If the area is a perfect square, compute

$$\mathcal{V} = \mathcal{A} \times \sqrt{\mathcal{A}}.$$

If not, compute

$$\mathcal{V}^2 = \mathcal{A}^2 \times \mathcal{A}.$$

If the quantities obtained are not integers, the result has the form of an integer with an additional fractional part.

Once again, as this supposes the computation of the area, which is obtained with an approximate ratio, even if the relation was correct, the answer obtained would have been an approximation. However, Bhāskara once again insists that Āryabhaṭa's rule is accurate, whereas the above mentioned "practical" relation is not.

E BAB.2.8

E.1 General rule

Āryabhaṭa gives a rule that can be summed up as follows: If $ABCD$ is an isoceles trapezium whose heights, AH, EG, BI are always equal to one another, as illustrated in Figure 12,

Figure 12: An isoceles trapezium

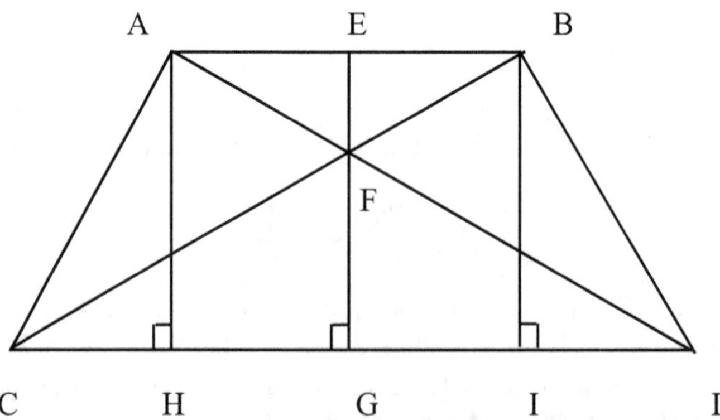

then:

$$EF = \frac{AB \times EG}{AB + CD}$$

$$FG = \frac{CD \times EG}{AB + CD}$$

and the area $\mathcal{A}$ is:

$$\mathcal{A} = EG \times \frac{(AB + CD)}{2}.$$

E.2   Description of the field

Bhāskara replaces, to a certain extent, Āryabhaṭa's terminology by his own. For instance, the uneven sides of an isoceles trapezium (AB and CD in Figure 12), are paraphrased by the commentator in the following way[17]:

> *ke te? pārśve| bhūr ekaṃ mukhami taram*
> What are those? The sides. One side is the earth, the other the face.

Bhāskara also explains the unusual technical term *svapātalekhā* (a line on its own falling)[18]:

> *svapātalekhā nāma antaḥkarṇayoḥ saṃpātasya*
> *bhūmukhamadhyasya cāntarālam*
> *Svapātalekhā* is the name of the inner space ⟨delimited by⟩ the intersection (*saṃpāta*) of the two interior ears and the middle of ⟨respectively⟩ the earth and the face.

He refers elliptically to these segments, by using the word *saṃpāta* (⟨the line(s) whose top is⟩ the intersection)[19], *saṃpātāgra* (⟨the line(s)⟩ whose tops is the intersection[20] and also with the compound *karṇāvalambakasaṃpāta* (⟨the lines whose tops are⟩ the intersection of the perpendicular and the ⟨interior⟩ ears[21]. We can note that previous translators of the Āryabhaṭīya seemed to have confused *svapāta* (a falling of one's own) and *saṃpāta* (an intersection). Thus Kaye[22] translates the compound *svapātalekhā* as if it was *saṃpātalekhā*: "the lines from the point of intersection". P. C. Sengupta[23] follows by giving the following translation, which is not literal: "the distance of the point of intersection of the diagonals from one

---

[17] [Shukla 1976; p.63]
[18] [Shukla 1976, p. 63]
[19] [Shukla 1976;p.63, lines 2 and 19]
[20] [Shukla 1976; Example 1, p.63]
[21] [Shukla 1976; p. 63, line 19]. Please refer also to the Glossary for the translations we have adopted of these terms.
[22] [Kaye 1908; p. 121]
[23] [Sengupta 1927; p.16]

of the parallel sides". Both Clark[24] and Shukla[25] seem to understand *svapāta* as relating to the orthogonality of the segments, and add the other understanding of the compound in parenthesis. In all cases, there is no ambiguity concerning the segments that this compound refers to.

The correspondence between Āryabhaṭa's technical terms for the sides of a trapezium and Bhāskara's are given in Table 6.

We can note here that if such segments are mediators for isoceles trapeziums this is not the case for any trapezium. When Bhāskara comments on the fact that they "fall in the middle" of the earth and the face, he thus restricts his description to the case of these trapeziums.

E.3   Bhāskara's interpretation

At the end of his general commentary on the verse, before the resolution of examples, Bhāskara gives an explanation, which may very well be a proof of the two rules given by Āryabhaṭa:

> *samyagādiṣṭena*[26] *likhite kṣetre svapātalekhāpramāṇaṃ*
> *trairāśikagaṇitena pratipādayitavyam|*
> *tathā trairāśikenaivobhayapārśve karṇāvalambakasampātānayanam|*

> The size of the 'lines on their own fallings' should be explained with the computation of a Rule of Three on a field drawn by ⟨a person⟩ properly instructed. Then, by means of just a Rule of Three with regard to the two sides which are a pair, the computation of ⟨the line whose top is⟩ the intersection of the diagonals and the perpendicular ⟨is made⟩.

Indeed, as illustrated in Figure 13, the triangles ABF and CFD are similar. Therefore

$$\frac{EF}{AB} = \frac{FG}{CD} = \frac{EG}{AB + CD}.$$

Such ratios are always given, in Bhāskara's commentary, as a Rule of Three. They are not stated explicitly here. One Sanskrit expression is rather difficult to understand here: *trairāśikenaivobhayapārśve* . Indeed we have translated it in this way: "by means of just a Rule of Three with regard to the two sides which are a pair". The compound *ubhayapārśve* should most usually be understood as: the side of both. We couldn't make much sense of all this...

---

[24][Clark 1930; 27]: "the perpendiculars (from the point where the two diagonals intersect) to the perpendicular sides".

[25][Sharma&Shukla 1976; p.42]: "the lengths of the perpendiculars on the base and the face (from the point of intersection of the diagonals)".

[26]Reading this instead of *samyagānadiṣṭena* of the printed edition.

Table 6: Names for the sides of a geometrical figure, as illustrated in Figure 12, given by Āryabhaṭa and Bhāskara.

Segments	Sanskrit names used by Āryabhaṭa	English Translation
AH, EG, BI	*āyāma*	height
AB, CD	*vistara, pārśva*	width, side
EF, FG	*svapātalekhā*	The 'lines on their own falling'

Segments	Sanskrit names used by Bhāskara	English Translation
AC, BD	*pārśva, karṇa*	side, ear
AB	*mukha, vadana*	face
CD	*bhū, bhūmi, dhātrī, vasudhā*	earth
AD, BC	*antaḥkarṇa, karṇa*	interior ears, diagonals
AH, EG, BI	*āyāma, vistara, dairghya*	height
AH, EG, BI	*avalambaka*	perpendicular
EF, FG	*svapātalekhā*	The 'lines on their own falling'
	sampāta	The ⟨lines whose top is⟩ the intersection ⟨of the interior ears⟩
	sampātāgra	The ⟨lines⟩ whose top is the intersection ⟨of the interior ears⟩
	karṇāvalambakasampāta	⟨The lines whose tops are⟩ the intersection of the perpendicular and the ⟨interior⟩ ears
HC, ID	*bhujā*	The base

Figure 13: Fields inside a trapezium

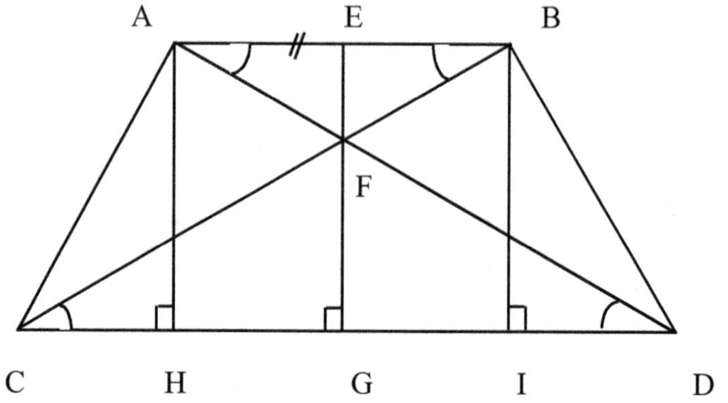

He further adds:

> *pūrvasūtreṇātra dvisamaviṣamatryaśrakṣetraphalaṃ darśayitavyam |*
> *vakṣyamāṇasūtreṇāntarāyatacaturaśrakṣetraphalaānayanam anena vā...*
> Here, with a previous rule (Ab.2.6.ab) the area of isoceles and uneven
> trilaterals should be shown. Or, with a rule which will be said (Ab.2.9.)
> the computation of the area   of the inner rectangular field ⟨should be
> made⟩

As illustrated in Figure 12, a trapezium can be seen as the sum of several triangles
(AFC, CFD, AFB and BFD) or as the sum of two right angle triangles (AHC
and BID) and a rectangle ($ABIH$).

Furthermore, Bhāskara distinguishes the case of isoceles trapeziums (which may
even have three equal sides as in Example 3) from the case of uneven trapeziums,
in the types of problems that may be solved by such a rule, the latter recquiring
a beforehand knowledge of the height of the trapezium. The procedure given by
Āryabhaṭa, in this case, only concerns the area.

In fact the part of the rule which computes the area of the trapezium is analyzed by
Bhāskara as being applicable to any quadrilateral. To state this property Bhāskara
needs to specify the terminology he is using. He therefore distinguishes what he
calls 'uneven quadrilaterals" (*viṣamacaturaśra*, i.e a non-isoceles trapezium), from
what is called with the same name in other treatises (i.e. any quadrilateral). To
do so he actually states a definition of what is trapezium:

> *atra ca yad upadiśyate tasya yāv avalambakau tau tulyasaṅkhyau|*
> The two perpendiculars of the ⟨field⟩ which is instructed here (in Ab.2.8)
> have the same value.

He then can write the above mentioned property:

atha yad gaṇitaśāstrāntaraupadiṣṭaviṣamacaturaśrakṣetra.m
yac cehaupādiśyate tayor dvayor api phalanirdeśo
py anenopadeśena śakyate [kartum]|
Now ⟨concerning⟩ that uneven-quadrilateral-field explained in a different treatise on mathematics and that ⟨field⟩ which is explained here (i.e
fields which have equal perpendiculars), the specification of the area of
these very two ⟨types of fields⟩ can be [made] with this instruction (i.e.
the one given in Ab.2.8.cd) as well.

E.4   Procedure followed in examples

E.4.1   Isoceles trapezium

Problem Knowing the sides, face and earth of an isoceles trapezium, find the two
lines on their own falling and its area.

Step 1 Find the height, considering an inner right angle triangle using Ab.2.17.ab.
As illustrated in Fig 12, considering triangle AHC or BID we have

$$CH = ID = \frac{CD - AB}{2},$$

and

$$EG^2 = AH^2 = AC^2 - HC^2,$$

or

$$EG^2 = BI^2 = BD^2 - ID^2.$$

In all examples here, the value found for the square of the height is a perfect
square.

Step 2 Compute according to Āryabhaṭa's rule the two segments of the height:

$$EF = \frac{AB \times EG}{AB + CD},$$

$$FG = \frac{CD \times EG}{AB + CD}.$$

Step 3 Compute according to Āryabhaṭa's rule the area of the trapezium:

$$A = EG \times \frac{(AB + CD)}{2}.$$

E.4.2 Uneven trapeziums

In this case, the height should already be given. Then both Step 2 and Step 3 of the previous procedure can be followed.

F BAB.2.9

F.1 Ab.2.9.ab

Āryabhaṭa gives the following general rule:

> *sarveṣāṃ kṣetrāṇāṃ prasādhya pārśve phalaṃ tadabhyāsaḥ|*
> For all fields, when one has acquired the two sides, the area is their product|

This is interpreted by Bhāskara in three ways: It is first read as giving a procedure to compute the area of rectangles. Then it is understood as a way of verifying the areas of the fields for which Āryabhaṭa has already given procedures that allow a computation of the area. Finally, it is read as a method to find the area of any field.

F.1.1 Procedure for the area of a rectangle

The area of the rectangle may be seen as a direct application of the method given by Āryabhaṭa here, as the area is a product of its width (*vistāra*) and length (*āyāma*). Bhāskara seems to admit that this is a very well-known fact. A verse quoted in the general commentary states:

> *vyaktaṃ phalam āyate yasmāt*
> since in rectangles the area is obvious

However, the first example of the commentary concerns rectangles.

F.1.2 Verifications

All the procedures given previously by Āryabhaṭa to compute the area of given fields can be seen as products of two quantities. Bhāskara re-reads these procedures as therefore producing the areas of rectangles having the same area as the initially computed field. He gives a name to this reasoning, it is called *pratyaykaraṇa*. Literally this word means "producing conviction". This we would translate as "proof" or "demonstration". However, historians of science seem to have all understood this word as meaning verification[27]. We have adopted the commonly used translation of this word.

We have discussed the nature of this reasoning as a verification in our thesis[28] giving two hypotheses on the nature of the reasoning elaborated here. The first hypothesis is that the reasoning considers the ability to find a rectangle with the same area as the field whose area is verified. This would interpret Ab.2.9.ab. as giving an essential property of plane geometry as conceived by Bhāskara: all fields can be transformed into a rectangle bearing the same area as the original field[29]. Another hypothesis would be to consider that a method was known starting with a given field to construct a rectangle with the same area. By then computing the area of such a rectangle, the area of the initial field would be verified. Bhāskara would then explicitly define arithmetically the link between the sides of the initial field and the newly constructed rectangle, explaining the validity of the procedure of construction.

These are hypotheses. We have here described the procedure followed formally, as they appear in the text.

a Verifying the area of trilaterals

a.1 equilaterals The idea is that the area of such triangles is equal to the rectangle whose sides are respectively the height and half the corresponding base, as illustrated in Figure 14.

Figure 14: An equilateral triangle and a rectangle with same areas

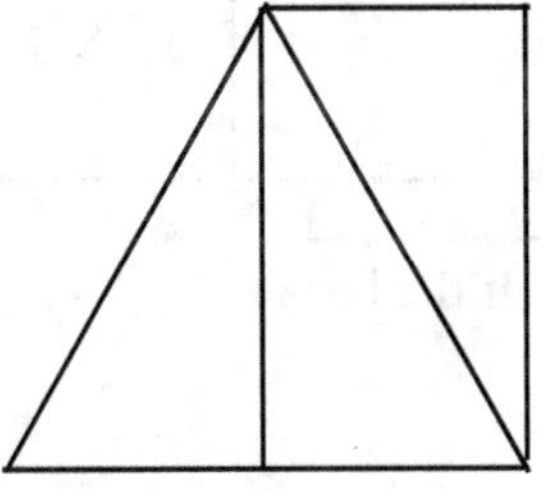

Problem Knowing the length of a side in an equilateral triangle, find the rectangle which has the same area and compute the area.

Step 1 Draw the triangle. Compute as described in BAB.2.6.ab. the height and half the base.

---

[27]See [Hayashi 1995; p. 72-75], who also analyses the use of the term in this text, and in non-mathematical texts. One can also see [Shukla 1976; intro p.liv]

[28]See [Keller 2000; I p. 104-127]

[29]This is exposed by T. Hayashi in [Hayashi 1995]

Step 2 Draw the corresponding rectangle. The area is the product of both.

The case of the isosceles is not treated. Bhāskara just adds:

> *evameva [dvi]sameṣu, viṣameṣu ca|*
>
> ⟨The computation⟩ is just like that in isosceles and uneven ⟨trilaterals also⟩.

b   Uneven trilaterals   Two methods are given. The first proceeds just as in the case of equilaterals, and therefore considers that any trilateral's area is equal to the area of the rectangle having for length and breadth respectively the height and half the base. We have given an illustration of this mathematical property in Figure 15, although no such drawing is in the text itself.

Figure 15: Any triangle has the same area as a rectangle whose sides are one height and half the corresponding base

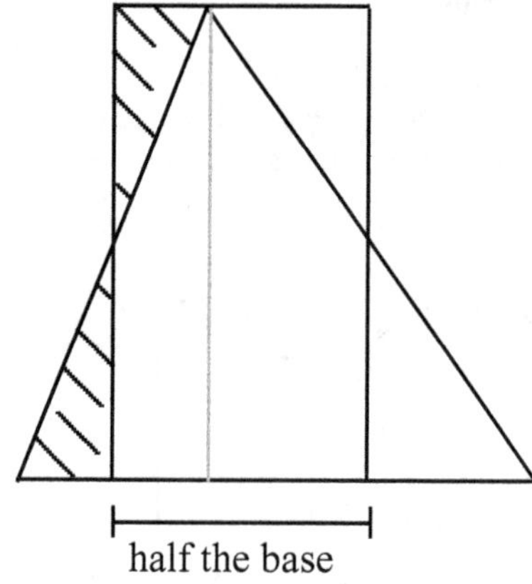

The second procedure is as follows, and is illustrated in Figure 16:

Step 1 Compute the sections of the base (BD, DC) created by the given height (AD) as described in BAB.2.6.ab.

Step 2 Compute the areas of the two rectangles (AEBD and AFDC), having drawn the corresponding figure. Halve the given areas.

Step 3 The area of the triangle is the sum of the half-areas of the rectangles.

Figure 16: Any triangle has the area of two half rectangles

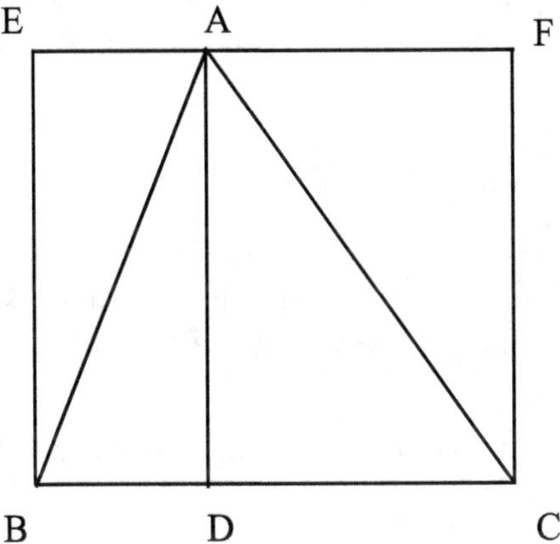

F.1.3   Circles

There is no illustration, but the following rule, is given:

> *vr̥ttakṣetre viṣkambhārdha vistāraḥ, paridhyardham āyāmaḥ,*
> *tad evāyatacaturaśrakṣetram*
>
> In a circular field, the semi-diameter is the width, half the circumference
> is the length, just that ⟨gives⟩ the rectangular field.

As we have noted before this rule seems a reinterpretation of the arithmetical rule
for computing the area of a circle as the product of two quantities, and, as an
arithmetical explicitation of the link between the segment and circumference of a
circle and the rectangle having the same area.

F.1.4   Trapeziums

Oddly, an isosceles trapezium is presented by Bhāskara as part of the group of mis-
cellaneous fields (*prakīrṇakṣetra*). This may be due to the fact that the trapezium,
as represented in a diagram, is considered here horizontally[30].

A trapezium has the same area as a rectangle having for sides, respectively its
perpendicular and half the sum of its parallel sides (or faces: *mukha* and *prati-
mukha*).

Problem Find the area of a trapezium whose two parallel sides and height is
known.

---

[30]We have discussed the sometimes implicit orientation of fields in [Keller 2000; I. p.228-230]

Step 1 Compute half the sum of the parallel sides.

Step 2 The area of the trapezium is equal to the area of the rectangle having for sides half the sum of the parallel sides and the height, therefore the area is their product.

F.1.5   A drum shaped, two dimensional figure

This field, illustrated in Figure 17, is characterized by a separation (*vyāsa*) or width (*vistāra*), corresponding to its smallest height (a) and its two parallel sides (*mukha*; b and c, which are equal in the only given example).

Figure 17: A two dimensional, drum-shaped field

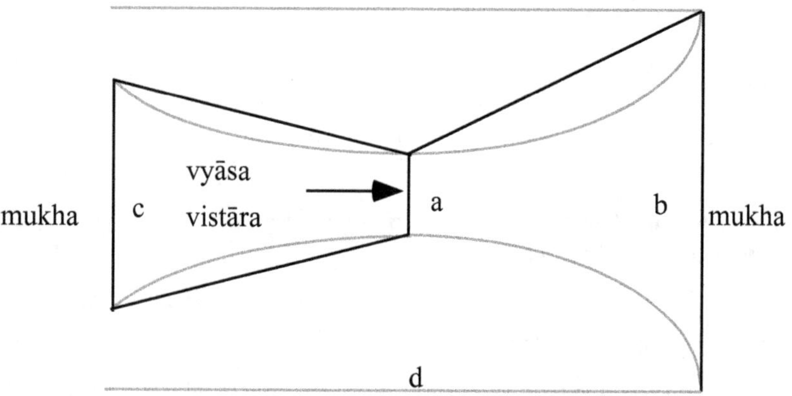

According to Bhāskara, the area of this field is the area of a rectangle having for sides respectively d and $\frac{\frac{b+c}{2}+a}{2}$ [31]. Therefore, its area $\mathcal{A}$ is

$$\mathcal{A} = d \times \frac{\frac{b+c}{2} + a}{2}.$$

This corresponds to the area of two trapeziums having a common parallel segment. In other words:

$$\mathcal{A} = \frac{d}{2} \times \frac{a+c}{2} + \frac{d}{2} \times \frac{a+b}{2}.$$

The diagram illustrating the solved example of this related text, in the edition, is a figure formed with two arcs (represented in filligrane in Figure 17): this may be

---

[31]The computation described by Bhāskara shows that the two sides can have different lengths. In the written example, even though the two sides are equal, Bhāskara writes: *mukhayoḥ samāsaḥ|* (The sum of the faces.) And afterwards he considers its half. He therefore computes: $\frac{2b}{2}$. In the computation of the following value he proceeds likewise.

Figure 18: A two-dimensional tusk field

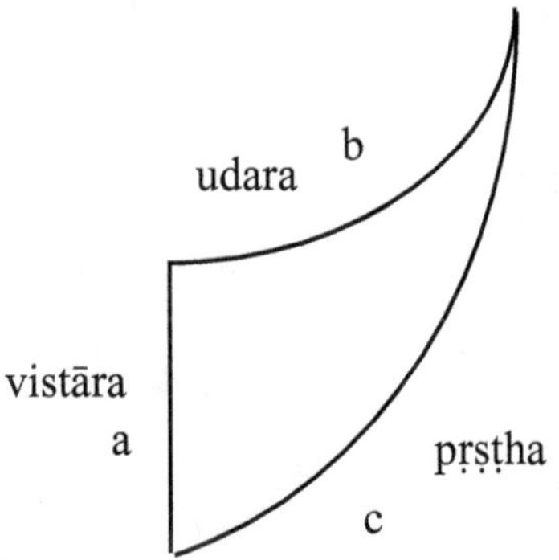

due to a deformation of the approximative straight lines often seen in the diagrams of palm-leaf manuscripts.

F.1.6 A two dimensional tusk-field

A tusk field, as illustrated in Figure 18, is characterized by a width (*vistāra, a*), a belly (*udara, b*) and a back (*pṛṣṭha, c*). Its area is considered to be equal to the rectangle whose sides are $(b+c)/2$ and $a/2$. Therefore the area, $\mathcal{A}$, of such a field is

$$\mathcal{A} = \frac{b+c}{2} \times \frac{a}{2}.$$

We do not know how this formula was found, but we can note that it presents an analogy with the formula giving the area of a circle. The area of such a field is known to have been studied in later mathematical texts. Some times it is considered as made of two arcs of a circle[32].

F.2 Ab.2.9.cd

Āryabhaṭa states in the second half of verse 9 that the chord that subtends an arc of 60 degrees is equal to the radius. This is illustrated in Figure 19.

F.2.1 Rāśis

A *rāśi*, as can be seen in Figure 19, is 1/12th of a circle, or 30 degrees. Bhāskara seems to consider the *arc* made of two *rāśis* as a field of its own. As we have stated in the Introduction, a circle is seen by Bhāskara not so much as a disk –

---

[32][Datta&Singh 1979; p.168 sqq]

Figure 19: The chord of a sixth part of the circumference, which is the chord subtending two *rāśis*, is equal to the radius

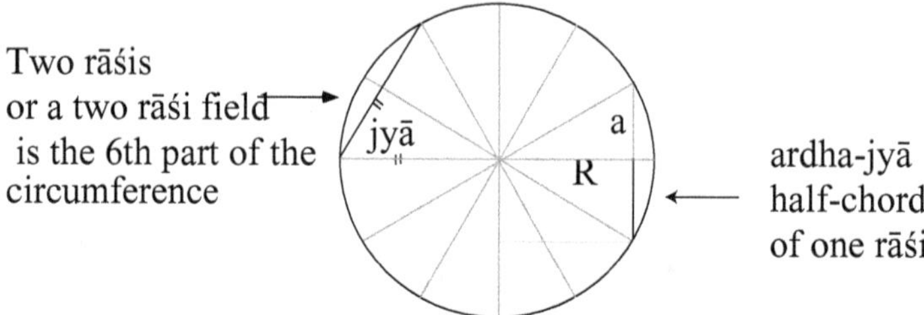

this is the idea of Prabhākara – then as the couple formed by a diameter and a circumference. In the same way, a two-*rāśi* field, even if the word "field" (*kṣetra*) conveys the idea of extension, would be restricted to the arc.

F.2.2 Half-chords

The vocabulary used in Bhāskara's commentary is confusing; but it makes sense in regard to the notion we use today of the sinus of an arc: If α is the measure of an arc measuring one *rāśi*, in a circle of radius R, half the chord of 2α is called by Bhāskara the half-chord of α. It corresponds precisely to $Rsin\alpha$, where an $Rsinus$ is the product of the sinus with the given radius. In other words:

$$\frac{chrd(2\alpha)}{2} = Rsin\alpha.$$

F.2.3 A pair of compasses

Bhāskara describes here, very briefly, a pair of compasses. The sentence where he does so, can be understood in various ways. For instance, the word *vartī* could refer to a piece of wood, a paint brush or some chalk. And the word *sita* could be a past participle (has been secured) or mean the color white. So that the same Sanskrit sentence

> *asmin ca viracitamukhadeśasitavartyaṅkurakarkaṭena ālikhite chedyake yat ṣaḍbhāgajyāyāḥ ardham tat rāśeḥ ardhajyā*|

can be read in at least five different ways. For instance as:

> And in this diagram, which is drawn with a compass with a white and sharp chalk (*sitavartyaṅkura*) fastened to the mouth-spot (*mukhadeśa*), that which is half of the chord of a sixth part is the half-chord of a *rāśi*.

or as

> And in this diagram, which is drawn with a compass with a secured
> (*sita*) and sharp paint brush (*vartyaṅkura*) fastened to the mouth-spot,
> etc.

Hence several images of compasses rise from this sentence. The interpretation we
have adopted rests upon Parameśvara's descriptions of a pair of compasses, which
we have discussed in the supplement for verse 13.

F.2.4 Fields within a circle

Figure 20: Fields seen inside a circle, whose circumference is divided in six equal
parts

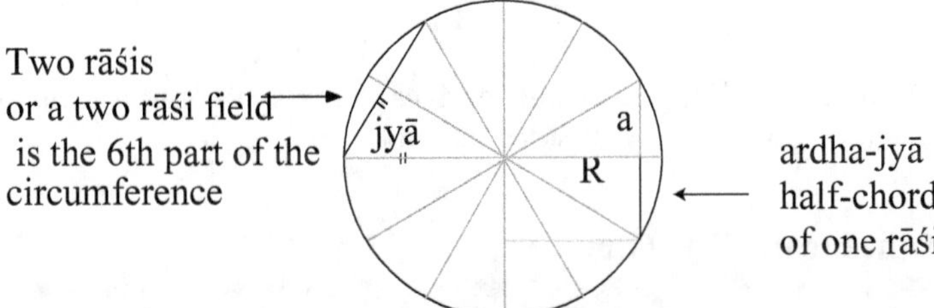

Bhāskara describes in the commentary several fields within a circle. The term
"*chedyaka*", which we have translated by "diagram" as it is used with this sense
in the *Mahābhaskarīya*, an astronomical treatise written by our commentator, is
only used in this commentary to refer to the figure whose drawing is described
in BAB.2.11[33]. Verse 11 of the chapter on mathematics is closely linked to this
one: it is the place where the application of such a relation will become clear in
Bhāskara's commentary.

G BAB.2.10

G.1 Āryabhaṭa's verse

Ab.2.10 relates a given diameter (measuring here 20000 units) to an approximate
circumference (62832). Bhāskara insists on the fact that an approximation of the
constant ratio linking the diameter of a circle ($2R$) and its circumference (C), which
we call π, is given here. A procedure to compute the diameter or circumference of

---

[33]Please see the supplement for this commentary for an illustration of the particular diagram
it may refer to.

any circle is deduced from this verse. It rests on a Rule of Three. The steps of the procedure to be applied in the case of the circle follow those of any Rule of Three, and are not exposed here.

With our notations, we can state the rule given by Bhāskara as follows. If a circle whose circumference is C and diameter is $2R$, then:

If $2R$ is known, approximately,

$$C = \frac{2R \times 62832}{20000}.$$

If C is known, approximately,

$$2R = \frac{C \times 20000}{62832}.$$

Or

$$\pi \simeq \frac{62832}{20000} = 3,1416^{34}.$$

According to Afzal Ahmad[35], this value derives from the computation of the perimeter of a regular polygon of 256 sides inscribed in a circle.

G.2 The "ten karaṇīs" theory

Bhāskara exposes in this part of his commentary another set of rules that may be ascribed to Jain authors. All are given in a dialect of Sanskrit. They are exposed in order to refute the first of these rules, which gives an alternative computation for the circumference of a circle. These rules have been discussed in [Shukla 1972].

The different steps of this refutation are given in the following subsection[36]. We will only unravel here the mathematical contents of each of these rules.

First rule   *vikkhaṃbhavaggad saguṇakaraṇī vaṭṭassa parirao hodi|*
    [viṣkambhavargadaśaguṇakaraṇī vṛttasya pariṇāho bhavati|]
    textbfThe *karaṇī* which is ten times the square of the diameter is
        the circumference of the circle|

To understand simply the mathematical idea of a *karaṇī*, one may consider it as a square root, although this is, to a certain extent, a heuristic transposition in our modern language. This rule can be formalized as follows: if C is the circumference of a circle, and $2R$ its diameter, this verse gives the computation

$$C = \sqrt{10.(2R)^2} \tag{1}$$

---

[34]For commentaries on approximations of π in India, see [Datta 1926], [Hayashi&Kusuba&Yano 1989], [Hayashi 1997b]

[35][Ahmad 1981]

[36]An analysis can be found in [Keller 2000; I p.120-126]

π is thus approximated as $\sqrt{10}$.

The Jain canonical works, known to us as preceding the time of Bhāskara, such as the *Sūryaprajñapti* (or *Sūryapaṇṇati*), use this value for π[37]. It is usually considered that such an approximation derives from the computation of the perimeter of a regular polygon with 12 sides, inscribed in a circle[38].

Second rule The second verse stated is:

> *ogāhūṇaṃ vikkhambhaṃ*[39] *egāheṇa saṃguṇaṃ kuryāt|*
> *caüguṇiassa tu mūlaṃ jīvā savvakhattāṇam||*
> *[avagāhonaṃ viṣkambham avagāhena*[40] *saṅguṇaṃ kuryāt|*
> *caturguṇitasya tu mūlaṃ sā jīvā sarvakṣetrāṇām||]*
> **The diameter decreased by the penetration should be multiplied**
> **by the penetration|**
> **Then the root of the product multiplied by four is the chord of all**
> **fields||**

The same verse, except for the last quarter, is given in verse 180 of the Jain work *Jyotiṣkaraṇḍaka*, an exposition in the line of the *Sūryaprajñapti*[41].

With the same notations as before, as illustrated in Figure 21, if a is the penetration (*avagāha*)[42], j (*jyā*), the chord, then

$$j = \sqrt{4(2R - a)a}. \tag{2}$$

This may be linked to the second part of verse 17 of the *Āryabhaṭīya*:

> **17cd. In a circular ⟨field⟩ (vṛtta), the square of the half chord,**
> **that is certainly the product of the arrows (śarasaṃvarga) of**
> **two bows||**

Let C be a circle of diameter AB and CDE a chord as illustrated in Figure 22, then we can understand the verse as

$$DE^2 = AD \times DB.$$

The two "bows" are thus the two arcs formed by CE, whose arrows are CD and DE.

If $j = 2DE$, and $DB = a$, so that $AD = 2R - a$, then we have

$$\left(\frac{j}{2}\right)^2 = (2R - a)a \Leftrightarrow j = \sqrt{4(2R - a)a}.$$

---

[37] See [Datta&Singh 1979; p. 152-154], [Hayashi 1997a; p. 12], [Sarasvati 1979; p. 62sqq]
[38] [Sarasvati 1979; p.65]
[39] The edition reads *vikkhambha*.
[40] The edition reads *avagāheṇa*.
[41] [Sarasvati 1979; p. 63, note 4]
[42] Or the "arrow" (*śara*), these two expressions refer to the same segments.

Figure 21: The field described in Bhāskaraś refutation

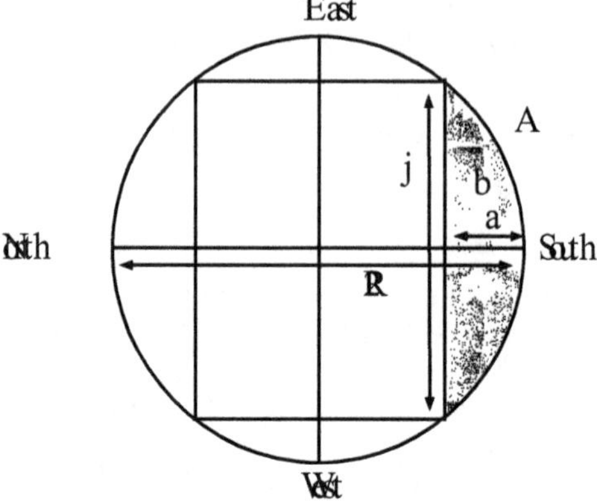

Figure 22: The figure illustrating the rule of the second half of Ab.2.17

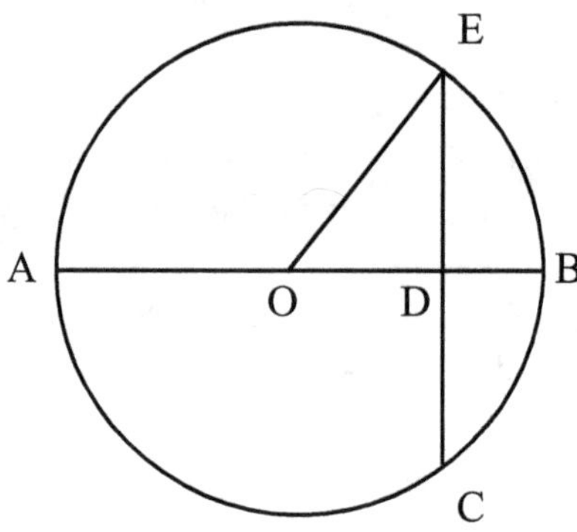

We can note that although this verse is quoted along with verses that are refuted, the fact that it can be seen as another formulation of Ab.2.17ab. shows that what is questioned by Bhāskara is not this procedure, which was probably considered correct, but precisely the value $\sqrt{10}$ used for the ratio of the diameter to the circumference of a circle.

Third rule   *isupāyagunā jīvā dasikarani bhaved viganiya padam|*
dhanupaṭṭa ammikhatte edaṃ karaṇaṃ tu āavvam‖

[iṣupādagunā jīvā daśakaraṇībhir[43] *bhaved viguṇya*[44] *phalam*[45]*|*
dhanuḥpaṭṭe śmin kṣetre etat karaṇaṃ tu jñātavyam‖]

**The chord with the quarter of the penetration as multiplier once
multiplied by ten karaṇīs will be the area|
In that field which is a strip like a bow, this procedure should be
known‖**

In other words, with the same notations as above, the area of a segment b of
a disk is

$$b = j \times \frac{a}{4} \times \sqrt{10}. \tag{3}$$

As before we do not know from where this computation was derived. We can
note that it is consistent with the computation of the area of a circle.

Fourth rule Then a rule to sum *karaṇīs* is stated:

aüvaṭṭi a dassakeṇa i mūlasamāsassamotthavat|
ovaṭṭanāyaguṇiyam karaṇisamāsam tu ṇāavvam‖
[apavartya ca daśakena hi mūlasamāsaḥ samottham yat|
apavartanāṅkaguṇitam karaṇisamāsam tu jñātavyam‖]
**When one has reduced ⟨the two karaṇīsto be summed⟩ by ten,
then, the sum of the roots ⟨of the results is taken⟩. That which
arises from the same ⟨sum⟩ (i.e. it is squared) is|
Multiplied by the digits of the reducer (i.e. ten), ⟨the result is a
karaṇī; in this way⟩ the sum of ⟨two⟩ karaṇīs should be known.‖**

K. S. Shukla gives the following formulae for this verse (in the introduction
p.lvi)

$$\sqrt{a} + \sqrt{b} = \sqrt{10 \left(\sqrt{\frac{a}{10}} + \sqrt{\frac{b}{10}} \right)^2}. \tag{4}$$

This is used when both $\frac{a}{10}$ and $\frac{b}{10}$ become perfect squares. So that the two
"$\sqrt{}$" symbols used over these quantities do not represent their irrationality
but a successful procedure of root extraction. Brahmagupta, a contemporary

---

[43]Although the plural instrumental ending makes sense Sanskritwise, it does not have any
parallel in the *prākṛta* verse.

[44]Likewise, if the substitution of the vowel *u* for the vowel *a* in *viganiya* makes sense mathe-
maticaly (the verb instead of meaning 'to compute', becomes'to multiply', it doesn't seem to be
based on any phonological evidence.

[45]Once again *phala* is more meaningful than *pada*, but isn't supported by phonology.

of Bhāskara, gives a rule to sum *karaṇīs* which is more general then this one, but follows the same idea[46].

Fifth rule   *jyāpādaśarārdhayutiḥ svaguṇā [daśasaṅguṇā karaṇyas tāḥ]|*
   The sum of a half arrow and ⟨its⟩ quarter-chord, multiplied to itself, [with ten as a multiplier, these are the karaṇīs ⟨that measure the back of the bow field⟩]|

In other words, with the same notation as before, considering that an arc p (*pṛṣṭha*) of a circle is computed knowing a chord j, and its arrow (or penetration) a

$$p = \sqrt{10\left(\frac{j}{4} + \frac{a}{2}\right)^2}. \tag{5}$$

We do not know from where this computation derives. It differs from those generally found in Jain canonical texts[47]. As this was rightly pointed out to me by Pr. Johannes Bronkhorst, this procedure is obviously false: if one adds the two complementary bow fields of one same chord (considering a and $2R - a$, one obtains the according circumference only if the chord is equal to the radius (e.g. if $j = 2R$).

G.3   Steps used to refute the "ten karaṇīs" theory

The global refutation is made of two separate refutations. The first one arrives at an impossibility of applying a given procedure – and the overall argument has to do with the expression of *karaṇīs* as numbers. In the second refutation, the result obtained is absurd.

The aim of the refutation is to discard $\sqrt{10}$ as an exact value of π. The second refutation, in fact, shows that it is an extremely rough approximation. Bhāskara proceeds by taking specific counter-examples. His reasoning rests not on the procedure quoted to compute the circumference of a circle, but on others that also use $\sqrt{10}$ as an approximation of π. (Namely those that we have transcribed as formulas (3) and (5)). He does not discuss the validity of these procedures as such, but seems to assume that, as they use the approximation he seeks to discard, this is the reason why they are faulty[48]. We present here the different steps that the two refutations take.

First refutation In the first refutation, Bhāskara attempts to compute the area of a circle as the sum of its interior fields. Though this is his program he does

---

[46]See [Hayashi 1997]

[47]See [Datta&Singh 1979; p.160sqq] and [Sarasvati 1979; p.63-64]

[48]For a more thorough analysis of the types of reasoning involved in the refutation see [Keller 2000; I p.120-126]

not, apparently, follow it to the end. He takes a specific case, in the form of a versified problem.

He then uses the procedure that we have transcribed as formula (3) to compute the areas of four bow-fields. They are obtained as *karaṇīs*. In order to sum the areas first of the bow fields, then of the interior rectangle he uses a rule that we have transcribed as formula (4). When trying to sum the areas of the bow fields, which amounts to 'ka.1210' (or $11\sqrt{10}$) and the area of the rectangle, which amounts to 'ka.2304' (or 48) he cannot obtain a simple number: in other words, he cannot write $11\sqrt{10} + 48$ as a single irrational quantity. Bhāskara states:

> *dhanuḥkṣetraphalasamāsarāśer asya ca karaṇisamāsakriyayā*
> *samasyamāne rāśyor asaṃkṣepatā|*
> When summing, with the method to sum *karaṇīs*, the quantity
> which is the sum of the areas of the bow fields and this (i.e, the
> area of the rectangle), both quantities are unsummable.

And this seems sufficient to show that an impossibility arises because of the use of $\sqrt{10}$. Takao Hayashi proposes to understand that as the area was considered to be the product of the circumference with the quarter of the diameter, the result obtained for the area of a circle should be written as one number and not as a non-reducible sum of *karaṇīs*.

However, if this was the case, wouldn't the procedure used to sum *karaṇīs* be what should have been under discussion?

We do not know if considering this procedure as part of the " ten *karaṇīs* theory", and thus considering it to derive from the use of this value for an approximation of π, it was to be discarded. We do not have an instance in another context in which Bhāskara attempts to sum *karaṇīs*.

Second refutation Bhāskara gives two counter examples for which the rule transcribed as formula (5) gives a value for the arc higher than that of its corresponding chord. This contradiction is commented upon by Bhāskara, twice, with some irony. The computation transcribed as formula (5) also uses $\sqrt{10}$ as a value for π, and therefore this procedure is seemingly refuted and not the one given for the circumference of the circle. Implicitly, Bhāskara assumes that the absurdity arises because the value for π is a very rough approximation.

Bhāskara concludes this refutation assuming that he has thus showed the impossibility of finding an exact procedure to compute the circumference of a circle knowing its diameter.

H   BAB.2.11

Bhāskara, in his commentaries on verses 11 and 12, aims at showing how the table of sine difference given by Āryabhaṭa in verse 12 of the first chapter of the *Āryabhaṭīya* is derived. This is not explicit in his commentary on verse 11, but becomes clear as we read BAB.2.12.

In this section, in a first part we will discuss Bhāskara's interpretation of Ab.2.11. In a second part we will explain the procedure he gives and in a third part we will discuss Bhāskara's remark concerning a chord equal to the arc it subtends.

H.1   Bhāskara's understanding of Ab.2.11.

In Ab.2.11. Āryabhaṭa just alludes to a geometrical situation (a circle whose circumference is first divided in quarters; trilaterals and quadrilaterals, related to arcs in a given quadrant...), in which half-chords should be computed, but he does not give any precise procedure. Both a geometrical context, namely a diagram, and a procedure followed within this diagram are supplied by the commentator.

H.1.1   "The quarter of the circumference of an even-circle"

The first quarter of verse 11 locates the procedure within a quarter of a circle (*samavṛttaparidhipāda*). The expression used in the verse to name a circle: *samavṛtta*, means "even circle". It is probably opposed to an "elongated circle" (*āyatavṛtta*), which is an ellipse. Bhāskara as he comments on the compound, indicates that what is considered is not the quarter of the disk but the quarter of the circumference. We will see how the procedure he provides uses several characteristics of the quarter of the circumference.

A *rāśi* is, in this case[49], a standard unit when considering a uniform subdivision of the circumference of a circle: it corresponds to 1/12th of the circumference, or 1/3rd of the quadrant.

Bhāskara states explicitly that the quadrant is convenient for it contains a whole number of *rāśis*, and that all the half-chords computed in one quarter are equal to those of other quarters.

H.1.2   "Trilaterals and Quadrilaterals": the diagram

The procedure Bhāskara gives may be understood as four sub-geometrical procedures used, within a diagram, to compute the length of a half-chord. This procedure will be described in a section below.

Bhāskara describes the construction of a diagram, very precisely, so that we can reconstruct it ourselves. Such a diagram is illustrated in Figure 23.

Figure 23: The diagram prescribed by Bhāskara

We note that in India the East (*pūrva*) is in front, the West (*paścima*) is behind, the North (*uttara*) is on the left and the South (*dakṣiṇa*) on the right. The cardinal directions are represented in the diagram of the printed edition of the commentary, but may not have been present in the manuscripts.

The procedure Bhāskara describes derives half-chords from right-angle triangles and a square that can be seen within the diagram he has prescribed. This is illustrated in Figure 24.

Figure 24: The trilaterals and the quadrilateral used by Bhāskara

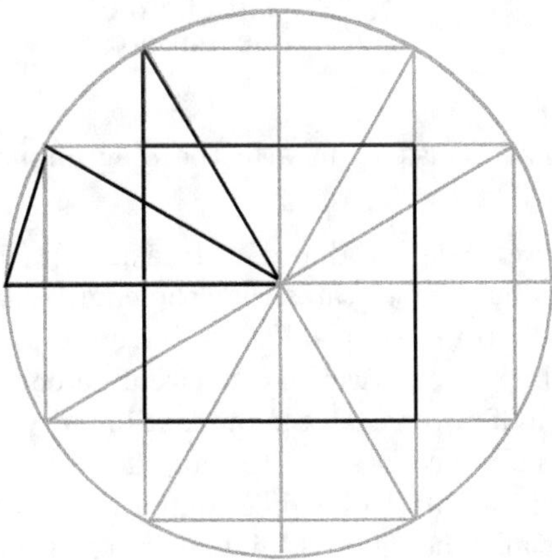

We note that rectangles appear also in this diagram: it is possible that Āryabhaṭa

---

[49]For the different meanings that *rāśi* can bear, please see the Glossary.

himself did not restrict his idea of "quadrilaterals" to the square considered by
Bhāskara.

H.1.3 Chords and half-chords.

This may be recalled here: Let there be a whole chord (*jyā*) subtending an arc
β. Half the chord subtending β is called the half-chord (*ardhajyā*) of the arc $\beta/2$.
This half-chord corresponds to the Rsinus (R times the sinus) of $\beta/2$.

This can be quite confusing as we read Bhāskara's commentary and is important
to bear in mind. In fact Bhāskara himself often omits the word *ardha* (half) when
he refers to a half-chord. In later works *jyā* or *jivā* alone name the half-chord[50].

A bow-field involves both a chord and a half-chord of a given arc, and also an
"arrow" (*śara*) which is ascribed to the arc of the half-chord. The arrow in the
case of a bow-field of two unit-arcs is illustrated in Figure 25.

Figure 25: A bow-field of two unit-arcs

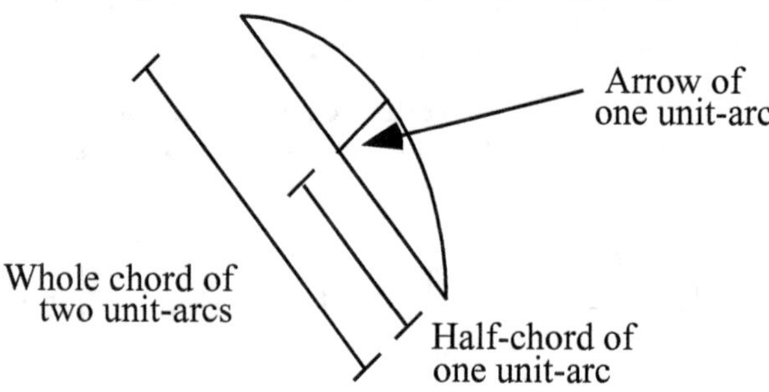

The arrow is a segment that Bhāskara uses in the diagrammatic procedure described below.

Chords and half-chords were first introduced in Bhāskara's commentary on the
second half of verse 9. This half-verse states that the chord subtending one sixth
of a circle is equal to the radius of the circle. He also introduces in this commentary
of verse the arc corresponding to one twelfth of a circumference, which is called a
rāśi. Thus with the second half of verse 9 we know that in any circle, the half-chord
of one twelfth of the circumference is equal to half the radius. The result given
by this verse is fundamental for Bhāskara's diagrammatic procedure, since it is on
the basis of this chord that all other chords (and their corresponding half-chords)
will be deduced.

---

[50]For remarks of later Sanskrit authors on the links between a chord and a half-chord see
[Datta&Singh 1983; p. 40]

H.1.4   Equal or even unit-arcs?

Bhāskara gives here a particular interpretation of the compound *samacāpa*, used in Ab.2.11: "*sama*" would be a reference to the fact that only half-chords of an even number of unit-arcs (*cāpa*) are to be produced by means of this procedure[51]:

> *jyāvibhāgena samavṛttaparidhau khaṇḍyamāne tribhujāc caturbhujāt*
> *ca kṣetrāt samacāpajyārdhāni niṣpadyante, na viṣamacāpajyārdhāni|*
> *tāni viśiṣṭāny eva parigṛhyante, dvicaturaṣṭaṣoḍaśadvātrimṡadityādini*
> *dviguṇauttarāṇi|*
> "the half-chords of an even ⟨number of⟩ unit-arcs" are produced "from
> a trilateral and a quadrilateral field", and not half-chords of an uneven
> ⟨number of⟩ unit-arcs.
> Just those particular ones, which are doubled successively, are understood:
> two, four, eight, sixteen, thirty two, etc...

Furthermore, Bhāskara glosses the word *tu* ("and") in order to add all the even arcs that this first interpretation omits:

> '*tu'śabdāt dvicatuṣṣaḍaṣṭadaśadvādaśacaturdaśādīni ca|*
> And, due to the word '*tu*' (and), two, four, six, eight, twelve, fourteen,
> etc... ⟨are understood.⟩

Two pieces of information are given to us in this part of the commentary.

First of all, we understand that Bhāskara assumes that the quadrant is divided by equal arcs.

Indeed, "equal arcs" could be another interpretation of the compound *sama* (equal)-*cāpa* (arc or unit-arc); this translation has been adopted by most of the translators of this verse with the exception of P.-S. Filliozat in [Filliozat 1988a]. Āryabhaṭa's idea may have been to insist that the arcs were equal, as for instance T. Hayashi has understood it in [Hayashi 1997], since the use of *sama* in the *Āryabhaṭīya* does not corroborate this interpretation of Bhāskara's as "even"[52].

As we have recalled above, the second half of verse 9 considers the twelfth-part of the circumference of a circle, which is called a *rāśi*. The twelfth-part of the circumference, or the third part of the quadrant, is the first, and most rough, subdivision (or partition, *vibhāga*) of the circumference that is considered by Bhāskara in the diagrammatic procedure.

What is called a "unit-arc" here is a given arc which produces a uniform subdivision of the circumference of a circle.

---

[51] [Shukla 1976; p. 77, line 15 sqq.]

[52] All occurrences of the word in the *Āryabhaṭīya* convey the meaning of "equal" or "uniformity".

In his commentary, Bhāskara, does not use the word *cāpa* (given by Āryabhaṭa) to name the unit-arcs considered but substitutes for it the word *kāṣṭha*. The unit-arcs called *kāṣṭha* considered in the procedure, are always an even subdivision of *rāśis* (i.e. 1/2, 1/4, 1/8th of a *rāśi*).

Table 7 gives the relations between *rāśis*, degrees and the number of unit-arcs considered.

Let us stress here that Bhāskara's interpretation of the compound leads him to understand that the diagrammatic procedure works only for half-chords of an even number of unit-arcs. As we will see in the next section, half-chords of an uneven number of unit-arcs can be derived from the diagrammatic procedure, and indeed they are, but when they are produced they stop the iteration of the process and indicate that a new procedure or a new field should be considered, in order to go further. Deriving a new half-chord with the half-chord of an uneven number of unit-arcs, with the given procedures, would indeed produce half-chords of a non-integer number of unit-arcs. This is probably why such a limitation is put forth.

The iterative aspect of the process is given by Āryabhaṭa with the expression *yatheṣṭāni* (as many as one desires).

H.1.5 "On the semi-diameter"

Bhāskara explains in three reasonings that complete one another how he understands the expression "the production of half-chords on the semi-diameter": First, the radius is fundamental because the trilaterals and the quadrilateral considered each have at least one side which is the radius. Secondly, the biggest value possible for the half-chord (the *R*sine) is the radius. Finally, the radius as the chord subtending one sixth of the circumference is the first numerical input that starts the procedure.

H.2 The steps of the diagrammatic procedure

As we have explained above, the procedure described by Bhāskara uses four different procedures, that each rest upon right-angle triangles and a square that can be drawn inside a circle. These are specific fields that are drawn along the uniform subdivision of the circumference into equal unit-arcs. In the procedures described by Bhāskara, even subdivisions of *rāśis* are considered. However the diagram whose construction is described in the commentary only considers a circumference subdivided by whole *rāśis*. Thus the diagram prescribed in the commentary is archetypical, it does not represent the effective triangles considered.

The four sub-procedures described in a diagram may be explained as follows[53]:

---

[53]The notations adopted are those used by Takao Hayashi in his article on Ab.2.12, [Hayashi 1997a]

Table 7: Number of *rāśis*, unit-arcs and degrees

For each application of the diagrammatic procedure, Bhāskara considers even subdivision of *rāśis*. This table gives the correspondence between the subdivision of *rāśis* considered, the three successive unit-arcs considered and the length in degrees of the arc considered.

rāśis	degrees	unit-arcs 1	unit-arcs 2	unit-arcs 3
1/8	3,75	-	-	1
1/4	7,5	-	1	2
3/8	11,25	-	-	3
1/2	15	1	2	4
5/8	18, 75	-	-	5
3/4	22,5	-	3	6
7/8	26, 25	-	-	7
1	30	2	4	8
9/8	33,75	-	-	9
5/4	37,5	-	5	10
11/8	41,25	-	-	11
3/2	45	3	6	12
13/8	48,75	-	-	13
7/4	52,5	-	7	14
15/8	56,25	-	-	15
2	60	4	8	16
17/8	63,75	-	-	17
9/4	67,5	-	9	18
19/8	71,25	-	-	19
5/2	75	5	10	20
21/8	78,75	-	-	21
11/4	82,5	-	11	22
23/8	86,25	-	-	23
3	90	6	12	24

Let 3×2^m, m being any integer, be the number of unit-arcs α in which a quadrant, of radius R, is divided (the quadrant then measures $3 \times 2^m \alpha$). A quadrant contains three *rāśis* (r), so that $r = 2^m \alpha$ (or $\alpha = \frac{r}{2^m}$). Let J_i be the *R*sine (*jyā*, R times the sine) of $i\alpha$, $0 < i \leq 3 \times 2^m$. This is illustrated in Figure 26.

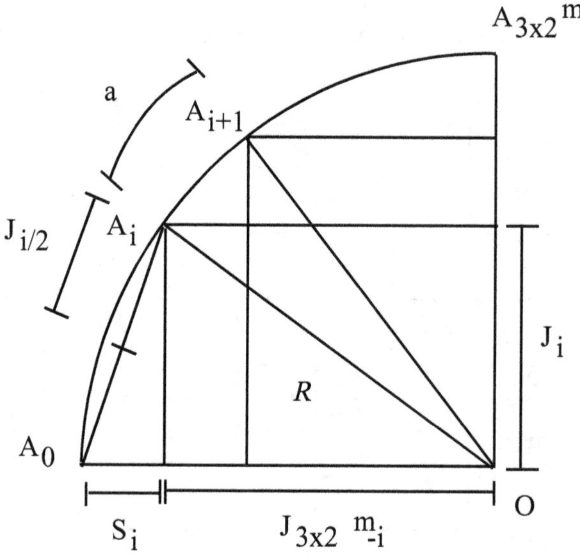

Figure 26: A quadrant with half-chords
The arc $A_i A_{i+1} = \alpha$, the arc $A_0 A_{3 \times 2^m} = 3 \times 2^m \alpha$.

procedure 1 Uses Ab.2.9.cd which states that the semi-diameter is equal to the whole chord of one sixth of the circumference (two *rāśis*). In other words, R is the whole chord of $2^{m+1}\alpha$.

Then

$$J_{2^m} = \frac{R}{2}.$$

This is the half-chord which can always be known and from which the iteration of the process may start.

For instance, in the first series of half-chords computed by Bhāskara, the unit-arc is half a *rāśi*, so that with our notation $m = 1$. Bhāskara shows that R is the whole chord of four unit-arcs, and therefore its half is the half-chord of two unit-arcs[54]:

---

[54][Shukla 1976, p.79, lines 7-8]

atrālekhye vyāsārdhatulyā caturṇāṃ kāṣṭhānāṃ [pūrṇa]jyā| tadardhaṃ dvikāṣṭhajyā|

In this drawing the [whole] chord of four unit-arcs is equal to the semi-diameter. Half of that is the ⟨half-⟩chord of two unit-arcs.

Procedure 2 With a known half-chord J_i, considering the right-angled triangle formed by R, J_i and $J_{3 \times 2^m - i}$, as illustrated in Figure 26, using Ab.2.17ab $J_{3 \times 2^m - i}$ is computed:

$$J_{3 \times 2^m - i} = \sqrt{R^2 - J_i^2}.$$

For example, in the first series of half-chords computed by Bhāskara, the unit-arc considered is half a *rāśi* ($m = 1$). From the half-chord of two unit-arcs (J_2) computed with procedure 1, the half-chord of four unit-arcs, illustrated in Figure 27, is computed according to the following geometrical reasoning[55]:

tadardhaṃ dvikāṣṭhajyā| sā ca 1719| eṣā bhujā, vyāsārdhaṃ karṇaḥ iti, bhujākarṇavargaviśeṣasya mūlam avalambakaḥ| saiva caturṇāṃ kāṣṭhānāṃ jyā| sā ca 2978|

Half of that is the ⟨half-⟩chord of two unit-arcs. And that is 1719. This is the the base, the semi-diameter is the hypotenuse, therefore the perpendicular is the root of the difference of the squares of the base and the hypotenuse. That exactly is the ⟨half-⟩chord of four unit-arcs. And that is 2978.

Another right-angled triangle considered is illustrated in Figure 28, as when Bhāskara, with the same unit-arc, computes the half-chord of five unit-arcs[56]:

eṣā bhujā, vyāsārdhaṃ karṇaḥ| bhujākarṇavargaviśeṣasya mūlaṃ koṭiḥ| sā ca pañcānāṃ kāṣṭhānāṃ jyā, sā ca 3321, viṣamatvād ato jyā notpadyante|

This (the half-chord of one unit-arc) is the base, the semi-diameter is the hypotenuse. The perpendicular is the root of the difference of the squares of the base and the hypotenuse. And that is the ⟨half-⟩ chord of five unit-arcs. And that is 3321. Because ⟨the number, 5, of unit-arcs⟩ is uneven, no ⟨half-⟩chords are produced from this.

As indicated in the last remark of the above quotation, if $J_{3 \times 2^m - i}$ is a half-chord of an uneven number of arcs (i.e $3 \times 2^m - i$ is uneven) then no new half-chord is derived with procedure 3. If this is not the case, procedure 3 is followed.

Procedure 3 With two known half-chords J_i and $J_{3 \times 2^m - i}$, $J_{\frac{i}{2}}$ is computed. A segment called the arrow (*śara*) of i unit-arcs (or the arrow of the half-chord

---

[55] [Shukla 1976; *opcit.* lines 8-9]
[56] *idem*, lines 11-13

Figure 27: A quadrant subdivided in half *rāśis*

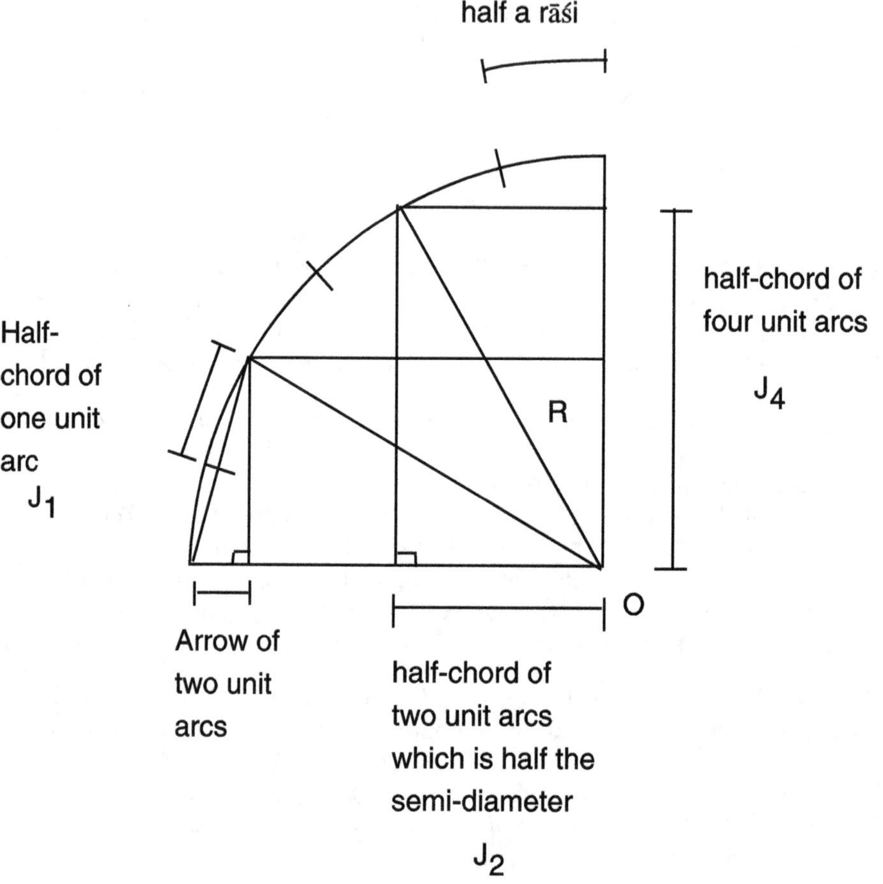

of i unit-arcs.), and noted here S_i, is considered. By definition:

$$S_i = R - J_{3 \times 2^m - i},$$

as illustrated in Figure 26. This computation considers the right-angled triangle formed by S_i, J_i and the whole chord of $i\alpha$, using Ab.2.17.ab.:

$$J_{\frac{i}{2}} = \frac{\sqrt{S_i^2 + J_i^2}}{2}.$$

Once $J_{\frac{i}{2}}$ is obtained, procedure 2 is used with $J_{\frac{i}{2}}$.

For instance, in the above exaxmple, Bhāskara, as illustrated in Figure 27, considers the right-angled triangle formed of the half-chord of two unit-arcs

Figure 28: Right-angled triangles in a circle

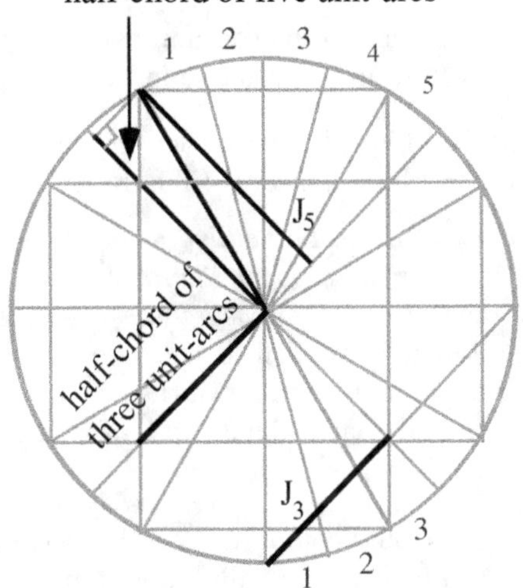

(J_2), the arrow of two unit-arcs ($S_2 = R - J_4$) and the whole chord of two unit-arcs, from which he deduces the half-chord of one unit-arc[57]:

> *etāṃ vyāsārdhād viśodhya śeṣaṃ dvikāṣṭhaśaraḥ,*
> *śaradvikāṣṭhajyāvargayogamūlaṃ karṇaḥ|*
> *saiva dvikāṣṭha[pūrṇa]jyā ca 1780|*
> *ardham asyāḥ kāṣṭhasyaikasya jyā, 890|*
> When one has subtracted this (i.e the half-chord of four unit-arcs) from the semi-diameter, the remainder is the arrow of ⟨the half-chord of⟩ two unit-arcs. The hypotenuse is the root of the sum of the squares of the arrow and the ⟨half-⟩chord of two unit-arcs. And that precisely is the [whole] chord of two unit-arcs, which is 1780. Half of that is the ⟨half-⟩chord of one unit-arc, 890.

From this half-chord of one unit-arc, with procedure 2 he deduces, as in the text quoted as an illustration in the description of procedure 1, the half-chord of five unit-arcs.

Procedure 4 Uses the fact that the diagonal of the square in the middle of the diagram, whose sides are equal to the semi-diameter (R), is the whole chord of three *rāśis* which is the whole chord of the quadrant itself. By using the "Pythagoras theorem" he can deduce the value of a half-chord.

---

[57][Shukla 1976; *idem* lines 9-11]

In other words

$$J_{3 \times 2^{m-1}} = \frac{\sqrt{2R^2}}{2}.$$

If $m = 1$, then the half-chord of an uneven number of unit-arcs is obtained and no new chord is derived from the value found for J_3.

When the unit-arc considered is half a *rāśi*, then $m = 1$, Bhāskara computes as follows, the half-chord of three unit-arcs (illustrated in Figure 28)[58]:

> *antaḥsamacaturaśrakṣetre vyāsārdhatulyā bāhavaḥ|*
> *tasya karṇo vyāsārdhayor vargayogamūlam|*
> *tac ca 4862| asyārdhaṃ trayāṇāṃ kāṣṭhānāṃ jyā| sā ca 2431|*
> In the interior equi-quadrilateral field the sides are equal to the semi-diameter. Its diagonal is the root of the sum of the squares of two semi-diameters. And that is 4862. Its half is the ⟨half-⟩chord of three unit arcs. And that is 2431.

The last relation shows – as one should compute $\sqrt{2}$ – that the square-roots given in this part of the commentary are systematically approximated.

This is illustrated in Table 8.

A similar table is given in [Hayashi 1997a], p. 402, where the line J_{*i} gives the approximate results according to the computation described in BAB.2.11. I do not find the same values as those given in this table (furthermore we have distinguished the approximate whole chords found from the half-chords that are deduced from them). This may be due to difference of approximations in the respective pocket calculators we have used to do these computations. Consequently the discrepancies of more than 0.5 do not always agree: although we both find discrepancies corresponding to the values of $J_6 \simeq 1215$, $J_7 \simeq 1520$ and $J_{16} \simeq 2978$. As explained by T. Hayashi in the above quoted article the three discrepancies observed may be explained by the fact that Bhāskara here is explaining how the table of sine difference given in Ab.1.12 was derived.

H.2.1 Additional Remarks

We can note that the restriction of the iteration of the procedure (from procedure 2 to procedure 3) to the half-chords of an even number of unit-arcs is probably due to the fact that it is always the Rsine of a whole number of unit-arcs that is considered. If i were uneven then $J_{\frac{i}{2}}$, computed by procedure 3, would give the half-chord of a non-integral number of unit-arcs.

The order in which the four procedures are applied in the diagrammatical procedure is illustrated in Table 9. Furthermore, Bhāskara seems to consider always an additional half-chord, since he systematically counts one more in the set of

---

[58]*idem,* lines 15-17.

Table 8: Bhāskara's given values and approximations

The results in bold indicate a discrepancy of more than 0.5 between the result stated in the commentary and the square-root obtained with an approximation of 10^{-2}. Arcs are considered in degrees. In his commentary on the following verse (BAB.2.12) Bhāskara comments on the process he uses when approximating quantities: for an integer obtained with an additional part smaller than a half the integer itself is used as an approximation; for an integer obtained with an additional part bigger than a half, the next integer is used as an approximation.

Arc in degrees	value given by Bh for the whole chord	Approx. value at a range of 10^{-2}	Half-chord (Rsin) derived	Given value of Rsin
7,5	450	449,94	Rsin3, 75	225
15	898	897,65	Rsin7,5	449
22,5	1342	**1340, 65**	Rsin11,25	671
30	1780	1779, 50	Rsin15	890
37,5	2210	2210, 15	Rsin18, 75	1105
45	26300	**2631, 31**	Rsin22,5	1215
52,5	3040	**3041,55**	Rsin26,25	1520
60	3438	-	Rsin30	1719
67,5	3820	**3821, 05**	Rsin33,75	1910
75	42876	4185, 85	Rsin37,5	2093
82,5	4534	4533,81	Rsin41, 25	2267
90	4862	4862, 07	Rsin45	2431

The discrepancies observed in the above table can be understood by the fact that the whole chord should be even: halved, it should produce a half-chord which is an integer.

Half-chord (Rsin)	Given value	Approximate value
Rsin48,75	2585	2584, 68
Rsin52, 5	2728	2727, 49
Rsin56, 25	2859	2858, 63
Rsin60	2978	**2977, 40**
Rsin63, 75	3084	3083, 74
Rsin67,5	3177	3176, 57
Rsin71, 25	3256	3255, 58
Rsin75	3321	3320, 80
Rsin78, 75	3372	3371, 88
Rsin82,5	3409	3408, 55
Rsin86,25	3177	3176,57

Table 9: Order of derivation of the half-chords in the diagrammatic procedure

procedure	Half-chord derived	unit-arc 1	unit-arc 2	unit-arc 3
1	J_{2^m}	J_2	J_4	J_8
2	$J_{2^{m+1}}$	J_4	J_8	J_{16}
3	$J_{2^{m-1}}$	J_1	J_2	J_4
2	$J_{5\times 2^{m-1}}$	J_5	J_{10}	J_{20}
3 applied with $J_{2^{m-1}}$	$J_{2^{m-2}}$	-	J_1	J_2
2	$J_{11\times 2^{m-2}}$	-	J_{11}	J_{22}
3 applied with $J_{2^{m-2}}$	$J_{2^{m-3}}$	-	-	J_1
2	$J_{23\times 2^{m-3}}$	-	-	J_{23}
3 applied with $J_{5\times 2^{m-1}}$	$J_{5\times 2^{m-2}}$	-	J_5	J_{10}
2	$J_{7\times 2^{m-2}}$	-	J_7	J_{14}
3 applied with $J_{5\times 2^{m-2}}$	$J_{5\times 2^{m-3}}$	-	-	J_5
p2	$J_{19\times 2^{m-3}}$	-	-	J_{19}
3 applied with $J_{11\times 2^{m-2}}$	$J_{11\times 2^{m-3}}$	-	-	J_{11}
2 gives $J_{13\times 2^{m-3}}$	-	-	-	J_{13}
3 applied with $J_{7\times 2^{m-2}}$	$J_{7\times 2^{m-3}}$	-	-	J_7
2	$J_{17\times 2^{m-3}}$	-	-	J_{17}
4	$J_{3\times 2^{m-1}}$	J_3	J_6	J_{12}
3	$J_{3\times 2^{m-2}}$	-	J_3	J_6
2	$J_{9\times 2^{m-2}}$	-	J_9	J_{18}
3 applied to $J_{3\times 2^{m-2}}$	$J_{3\times 2^{m-3}}$	-	-	J_3
2	$J_{21\times 2^{m-3}}$	-	-	J_{21}
3 applied with $J_{9\times 2^{m-2}}$	$J_{9\times 2^{m-3}}$	-	-	J_9
2	$J_{15\times 2^{m-3}}$	-	-	J_{15}

half-chords obtained. This most probably is the half-chord which has for length the radius.

If we look at the geometrical aspect of the procedures applied, and especially at what the balancing between procedure 2 and procedure 3 effectively does, we can notice that procedure 2 always produces a segment orthogonal to the one it derives from. Procedure 3 produces the segment of a hypotenuse from which another orthogonal side may be produced.

The graphic aspect of the process is illustrated in Figure 29, in the case where the unit-arc corresponds to half a *rāśi*; and in Figure 30, in the case where the unit-arc corresponds to a quarter of a *rāśi*.

Figure 29: Geometrical representation of the half-chords derived for a unit-arc equal to half a *rāśi*

The number between () indicates the order in which the half-chord is derived.

procedure	half-chord derived	with unit-arc 1
1	J_{2m}	J_2
2	J_{2m+1}	J_4
3	J_{2m-1}	J_1
2	$J_{5\times 2^{m-1}}$	J_5
4	$J_{3\times 2^{m-1}}$	J_3

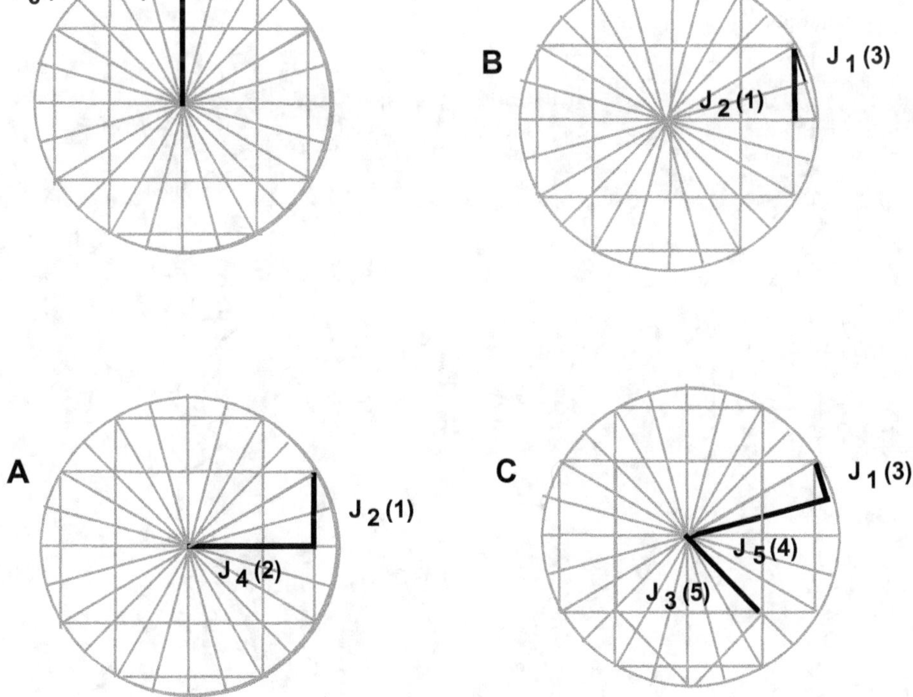

Figure 30: Geometrical representation of the half-chords derived, when the unit-arc is a quarter of a *rāśi*
The number between () indicates the order in which the half-chord is derived.

procedure	Half-chord derived	with unit-arc 2
1	J_{2^m}	J_4
2	$J_{2^{m+1}}$	J_8
3	$J_{2^{m-1}}$	J_2
2	$J_{5\times 2^{m-1}}$	J_{10}
3 applied with $J_{2^{m-1}}$	$J_{2^{m-2}}$	J_1
2	$J_{11\times 2^{m-2}}$	J_{11}
3 applied with $J_{5\times 2^{m-1}}$	$J_{5\times 2^{m-2}}$	J_5
2	$J_{7\times 2^{m-2}}$	J_7
4	$J_{3\times 2^{m-1}}$	J_6
3	$J_{3\times 2^{m-2}}$	J_3
2	$J_{9\times 2^{m-2}}$	J_9

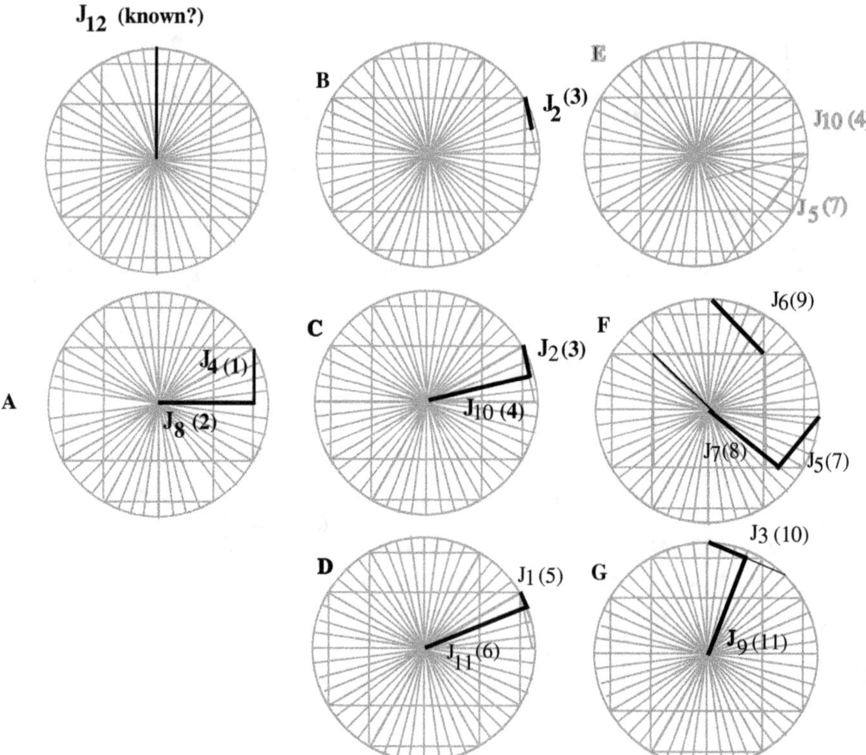

H.3   A chord of the same length as the arc it subtends

Bhāskara states here his dissension with another scholar, Prabhākara, concerning the existence of an arc having the same length as the chord it subtends. He quotes a verse:

> **Because of its sphericity (golaka-śarīra) ⟨a sphere⟩ touches the earth with the hundredth part of its circumference |**

In Lalla's *Śiṣyādhīvṛddhida* (VIIIth or early IXth century according to Pingree, beginning of Xth century and before the middle of the XIth century according to Billard) and in the *Siddhāntaśiromaṇi* by Bhāskara II (1150 A.D.) similar opinions are stated. We can note that the sphere, and the great circle of such a sphere, seem here to be confused or at least collected in the same idea.

Bhāskara, however, states that this arc, which can be assimilated to its chord, is the 96th part of the circumference. The 96th part of the circumference corresponds to the unit-arc measuring one eighth of a *rāśi*. The half-chord of such an arc is computed by Bhāskara as measuring 225. Now according to the procedures described in BAB.2.10 giving the ratio of an arc to its subtending chord, we can show that because of the type of approximations used for extracting the square root, both the whole chord and the arc measure 225 as well.

225 appears thus as the smallest unit for which computations of arcs and chords could be carried out by Bhāskara. Āryabhaṭa's sine difference table starts with the value 225.

I   BAB.2.12

In Ab.2.12, Āryabhaṭa gives a method to compute a series of Rsine differences. Several understandings of this verse have been discussed by historians of mathematics, following different commentators of Āryabhaṭa. They have all been listed in [Hayashi 1997a; p.398-399]. Takao Hayashi himself gives a new interpretation, in this article (p. 399 sqq), of this verse based on Nīlakaṇṭha's (born 1444) interpretation. The particularity of Bhāskara's (mis)understanding is – beyond a specific grammatical and semantic analysis of the rule which brings forth a specific procedure – to link it with the preceding verse. This analysis disqualifies the rule in his eyes. This interpretation, however, is ascribed by Bhāskara to Prabhākara, a scholar of whom we do not have any work but whose interpretations of Āryabhaṭa's rules are often discussed in this commentary.

We will not discuss here, for mere lack of time, how several such interpretations can arise from Āryabhaṭa's verse[59]. Such a thread, as it would highlight the interpretation a commentary ascribes to a given rule, would be of great interest.

---

[59]Indeed, this would involve reading the Sanskrit commentary of each author and analyzing in

In this section, we will first explain Bhāskara's understanding of this verse, indicating here and there and in no way exhaustively, alternative interpretations given by other commentators. In a second part we will give the different steps of the procedure he prescribes, and the method of approximation he uses. In a brief last section we will comment on the last sentence of this commentary, which deals with Rversed sines.

I.1 A specific interpretation of the rule

This way we have translated Bhāskara/Prabhākara's understanding of Ab.2.12:

> *prathamāc cāpajyārdhāt yair ūnaṃ khaṇḍitaṃ dvitīyārdham|*
> *tatprathamajyāardhāṃśais tais tair ūnāni śeṣāni‖*
> The segmented second half-⟨chord⟩ is smaller than the first half-chord
>   of a ⟨unit⟩ arc by certain ⟨amounts⟩ |
> The remaining ⟨segmented half-chords⟩ are smaller ⟨than the first half-
>   chord, successively⟩ by those ⟨amounts⟩ and by fractions of the first
>   half-chord accumulated.‖

I.1.1 Segmented half-chord of unit-arcs

What Bhāskara calls a "segmented half-chord of unit arcs" (*cāpajyārdhaccheda* or *cāpajyārdhāṃśa*) is the object of the computation here, a difference of Rsine. Indeed the difference of two Rsine, can be seen, geometrically, as a segment of the largest of the two half-chords considered. This is illustrated in Figure 31.

In this verse, as in Bhāskara's commentary, the half-chords form an ordered set: the half-chord of one unit-arc is called "the first half-chord ⟨of a unit-arc⟩", the half-chord of two unit-arcs is called "the second half-chord ⟨of two unit-arcs⟩" and so on. Numerically, the set of half-chords considered is the one that was derived in BAB.2.11 for a unit-arc measuring 1/8th of a *rāśi*.

"Segmented half-chord" is Bhāskara's interpretation of one expression of Āryabhaṭa's verse: *khaṇḍitaṃ dvitīyārdham* (the segmented second half-chord), that he glosses as follows:

> *khaṇḍitaṃ dvitīyārdham, khaṇḍitaṃ pūrvāryābhihitachedyakavidhinā chinnaṃ*
> "The segmented second half ⟨chord⟩", ⟨it⟩ is segmented, ⟨in other words⟩
> the second half-chord of ⟨unit⟩ arc is cut (*chinna*) by means of the dia-
> grammatical rule (*chedyakavidhi*) told in the previous *āryā* ⟨verse⟩.

The use of *chinna* here might be a pun. *Chinna* obviously glosses *khaṇḍita*, both can have the meaning of "divided", "segmented", "cut". Only in BAB.2.11 no

what way, syntactically, semantically, and mathematically, this interpretation has been derived. Takao Hayashi, in the above mentioned article, just presents what the final reading amounts to, and provides a mathematical analysis of them.

Figure 31: K_{i+1} appears as a "section" or segment of J_{i+1}

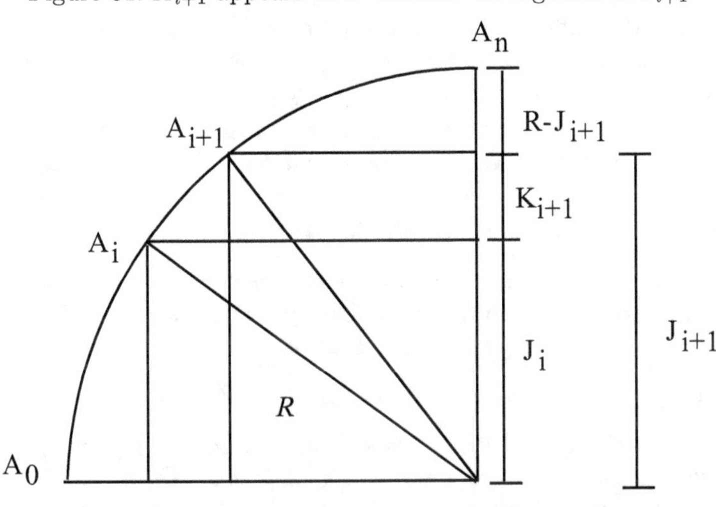

"segmented" half-chord, i.e. no sine difference is obtained. But the word translated as "diagram", *chedyaka*, uses the same verbal root, *ChID-*, as *chinna*. So that we can understand the use of this word as both referring to the fact that the sine difference using the second half-chord is obtained with a diagrammatic method (by taking the difference of the half-chords obtained by the procedure described in BAB.2.11), and that it is a segment of a half-chord.

The first half of the verse, as Bhāskara understands it, therefore compares the first half-chord with the difference between the first half-chord and the second half-chord.

In other words, using the same notations as those used in our supplement for BAB.2.11: Let 3×2^m be the number of ⟨unit⟩ arcs, α, a quadrant is divided in, J_i is the Rsine of αi, ($0 \leq i \leq 3 \times 2^m$). And let for $i > 1$, $K_i = J_i - J_{i-1}$ be the Rsine differences (*khaṇḍitam ardhajyām*). (This is illustrated in Figure 31)

Bhāskara therefore understands the first half of the verse, as concerning the difference $J_1 - K_2$.

We can note here that Bhāskara understands the expression *cāpajyārdha* in the first quarter of the verse as meaning "the half-chord of one unit-arc". Nīlakaṇṭha, with a different interpretation of the same compound, understands it, in T. Hayashi words, as meaning "the first half-chord, which is ⟨approximately equated to⟩ the ⟨corresponding⟩ arc (α)". The first half-chord considered in Āryabhaṭa's table is 225, a value that we have noted in BAB.2.11 corresponds to what Bhāskara calls "a chord equal to its arc".

I.1.2 "Certain Amounts"

Bhāskara considers this difference between the first half-chord (which is the half-chord of one unit arc), and the difference between the first half-chord and the second half-chord ($J_1 - K_2$) always in a plural form.

This arises from his interpretation of the instrumental plural relative pronoun of the first half of the verse: *yais*. We have translated it as: "(is smaller) by certain ⟨amounts⟩".

Glossing this term, Bhāskara writes (p. 83, line14):

> *yair ūnaṃ yāvadbhir aṃśair ūnam aprāptasadṛśam*
> (The second partial half chord) "is smaller by certain ⟨amounts⟩", ⟨it⟩
> is smaller, that is shorter (than the first half-chord), by certain parts.

But when he computes this difference (p. 84, line 4), he writes:

> *prathamaṃ cāpajyārdham idaṃ chedyakena niṣpannam 225| dvitīyaṃ cāpajyārdhac chedam 224| etat prathamacāpajyārdhād ekenonam|*
> This first half-chord of ⟨unit⟩ arc produced with a diagram is 225. The second partial half-chord of ⟨unit⟩ arcs is 224. This is smaller than the first half-chord of ⟨unit⟩ arc by one.

So that the "parts" or "certain ⟨amounts⟩" given in the plural form, amount, in this case, to one unit. Evidently here, Bhāskara's interpretation is not consistent with what he computes.

This plural form may be, however, understood less literally: It can be seen as an elliptic formulation used by Bhāskara to indicate that the difference of the two first half-chords ($J_2 - J_1$), should be considered in a plural form. Indeed, the idea of a "plurality of amounts", a way of indicating a number which is higher than one appears p. 83, line 16:

> *yāvadbhiḥ prathamacāpajyārdhād dvitīyacāpajyārdham ūnaṃ tāvantas taiḥ parigṛhyante*
> . . . they understand so many ⟨amounts⟩, by means of which the second half-chord of ⟨unit⟩ arcs is less than the first half-chord of a ⟨unit⟩ arc.

We can also understand it as expressing a general case: it is only in the table computed here that $J_1 - K_2$ is unity.

If we do not accept these hypotheses, we will then conclude that this plural form is certainly due to Bhāskara's misunderstanding of Āryabhaṭa's rule. In fact the relative plural pronoun most probably is to be ascribed to the second half verse.

I.1.3 "fractions accumulated"

The second half of the verse states, as Bhāskara understands it, that this first difference $J_1 - K_2$, and "fractions of the first half-chord accumulated" (*prathamacāpa-*

jyārdhāṃśa) when subtracted from the first half-chord give the remaining Rsine differences.

Bhāskara is quite elusive in his general commentary on what these fractions are. He states that the first half-chord is their denominator:

> *prathamajyārdhāṃśāś ca prathamajyārdhena bhāgaṃ hṛtvā labdhā yathā*
> *pañcāṃśah, ṣaḍaṃśah*
> And a fraction of the first half-chord is what has been obtained when one has divided by the first half-chord, just like "a fraction of five" (one fifth) and "a fraction of six" (one sixth).

He also indicates that their is an "accumulation" of these fractions: this should be understood as meaning that they are added. These are the only two elements that are explained by Bhāskara in his general commentary. It is by following the effective computation of the first five Rsine differences that we get a clear idea of the computation he bears in mind, as we will see in the next section.

I.2 Understanding the procedure

Now, with the same notations as before, let 3×2^m be the number of $\langle$unit$\rangle$ arcs, α, a quadrant is divided in, J_i is the Rsine of αi, ($0 \leq i \leq 3 \times 2^m$). And let for $i > 1$, $K_i = J_i - J_{i-1}$ be the Rsine differences (*khaṇḍitārdhajyā*). The computation of a given K_{i+1}, knowing K_i may be understood as follows:

Step 1 "The segmented second half-$\langle$chord$\rangle$ is smaller than the first half-chord of a $\langle$unit$\rangle$ arc by certain $\langle$amounts$\rangle$": Consider $J_1 - K_2$.

Assuming

$$J_1 = 225,$$

$$K_2 = J_2 - J_1 = 224.$$

Then

$$J_1 - K_2 = 1.$$

We have noted above that even though this difference is considered to be one, it is always referred to in a plural form. This may indicate that this interpretation of Āryabhaṭa's verse has a flaw. It may also be an elliptic formulation, where the plural, in fact refers to $J_2 - J_1$, or an indication that the computation considered here is a particular one: considered in all its generality, $J_1 - K_2$ can be higher than 1.

Step 2 Compute a "fraction of the first half-chord", that is the quotient of the sum of the first half-chord and of all the partial half-chords already computed (all the $K_j, 2 < j \leq i$) with the first half-chord. In other words, compute

$$\frac{J_1 + \sum_{n=2}^{i} K_n}{J_1}.$$

If the non-integer part of the quotient is greater than a half, approximate the quotient by adding 1.

This step is not given in Āryabhaṭa's verse, nor in Bhāskara's general commentary. When computing, for example, the 4th sine difference, knowing that $K_3 = 222$, Bhāskara writes:

> *trayāṇāṃ saṃyogaḥ 671| asya prathamacāpajyārdhena*
> *bhāgalabdham ardhādhikena trīṇi rūpāṇi*
> The sum of the three ⟨partial half-chords⟩ is 671. The division of that with the first half-chord of a ⟨unit⟩ arc ⟨is made⟩, the quotient, because it is greater than one half, is three unities.

In other words $J_1 + K_2 + K_3$ is considered ($J_1 = K_1$ being the short-cut adopted for the brackets.) We then have

$$J_1 + K_2 + K_3 = 225 + 224 + 222 = 671.$$

Then a process of approximation is clearly described. In this case, the quotient considered is

$$K_4 = \frac{J_1 + K_2 + K_3}{J_1} = \frac{671}{225} = 2 + \frac{221}{225}.$$

As $\frac{221}{225} > \frac{1}{2}$, the whole quotient is approximately considered to be equal to 3.

Step 3 "The remaining ⟨segmented half-chords⟩ are smaller ⟨than the first half-chord, successively⟩ by those ⟨amounts⟩ and by fractions of the first half-chord accumulated." In other words:

$$K_{i+1} = J_1 - \left\{ (J_1 - K_2) - \sum_{j=2}^{i} \frac{J_1 + \sum_{n=2}^{j} K_n}{J_1} \right\}.$$

Because

$$J_1 + \sum_{n=2}^{j} K_n = J_j,$$

we would have

$$K_{i+1} = J_1 - \left\{ (J_1 - K_2) - \sum_{j=2}^{i} \frac{J_j}{J_1} \right\}$$

as stated in [Hayashi 1997a; note 5 p. 399].

For example, when computing the fourth sine difference, Bhāskara writes:

> *taiḥ pūrvalabdhaiś ca tribhir ūnaṃ prathamacāpajyārdhaṃ*
> *caturthajyārdhaṃ bhavati| tac ca 219|*
> The fourth ⟨partial⟩ half-chord is smaller than the first half-chord of ⟨unit⟩ arc, by these ⟨three⟩ and by the previously obtained fractions, and that is 219.

In a previous computation, the approximate quotient was given

$$\frac{J_1 + K_2}{J_1} \simeq 2.$$

So that here

$$K_4 = J_1 - (J_1 - K_2) - (\frac{J_1 + K_2}{J_1}) - (\frac{J_1 + K_2 + K_3}{J_1}).$$

Or numerically:

$$K_4 = 225 - 1 - 2 - 3 = 219.$$

This process is reiterated in order to obtain all K_i's for $1 < i \le 3 \times 2^m$.

For a mathematical analysis of this computation, please see [Hayashi 1997a].

I.3   Rversed sine

Bhāskara ends this verse by declaring:

> *etā evotkramenāntyād ārabhyotkramajyāḥ*
> These ⟨partial half-chords, added⟩ in the reverse order beginning from
> the last, are the *utkramajyā* (Rversed sine).

Although he does not elaborate, we can notice that since $R = J_{3 \times 2^m}$, the last
sine difference corresponds to $R - J_{3 \times 2^m - 1}$ which is the Rversed sine of the arc
$\alpha(3 \times 2^m - 1)$. By summing the differences of the half-chord in reverse order, we
obtain in this way successively $R - J_{3 \times 2^m - 2}$, $R - J_{3 \times 2^m - 3}$ etc. This (the segment
$R - J_i$) is what bears the name *utkramajyā* or Rversed sine. This segment is
often used by Bhāskara with other names: it is the arrow (*śara*) of the half-chord
of $\alpha(3 \times 2^m - 1)$ in BAB.2.11 for instance, or the penetration (*avagāhin*) when
considering two intersecting circles, as we can see in the commentary on verse 18.

J   BAB.2.13

J.1   What Bhāskara says of compasses

A pair of compasses appears among the tools quoted by Āryabhaṭa in this verse.
Āryabhaṭa calls compasses a *bhrama* "a rolling ⟨object⟩". Bhāskara calls it a
karkaṭa or *karkaṭaka*, literally a "crab". In his commentary on verse 13 Bhāskara
gives only a brief explanation of this object[60]:

---

[60][Shukla 1976; p.85]

> *bhramaśabdena karkaṭakaḥ parigṛhyate| tena karkaṭakena samavṛttaṃ*
> *kṣetraṃ parilekhāpramāṇena parimīyate|*
> With the word *bhrama* a pair of compasses (*karkaṭa*) is understood.
> With that pair of compasses an evenly circular field is delimited by the
> size of the out-line (*parilekhā*).

Elsewhere he is slightly more specific. Thus in his commentary on the latter half
of verse 9 of the chapter on mathematics, he writes[61]:

> *asmin ca viracitamukhadeśasitavartyaṅkurakarkaṭena ālikhite chedyake...*
> And in this diagram, which is drawn with a compass (*karkaṭa*) for
> which a sharp stick (*vartyaṅkura*) secured (*sita*) at the mouth spot
> (*mukhadeśa*) has been arranged....

As we have noted in our supplement for verse 9, according to the meanings we
give to *vartī* (or *vartikā*; usually the wick of a lamp, a paint-brush or chalk) and
to *sita* (has been fastened, white color), different readings of this description are
possible, and hence different images of compasses appear. We also do not know
what is a compass' "mouth spot" (*mukhadeśa*). The same difficulties arise when
we read the short description in Bhāskara's commentary on verse 11[62]:

> *tathā ca paridhiniṣpannaṃ kṣetraṃ karkaṭakena viracitavartikāmukhena*
> *likhyate*
> And thus a field produced by a circumference is drawn with a pair of
> compasses whose opening (*mukha*) has a sharpened stick (*viracitavar-*
> *tikā*).

We have adopted the improbable reading of *vartī* (or *vartikā* that we have read
as a synonym of the first) as "stick" by accepting Parameśvara's interpretation of
the compound *vartikāṅkura*.

J.2 Parameśvara's descriptions of a pair of compasses

Parameśvara is a well known as a prolific astronomical commentator of the XVth
century[63]. He wrote commentaries on Bhāskara II's works as well as on the *Ārya-*
bhaṭīya[64]. He also wrote a direct and a super commentary on Bhāskara I's *Mahā-*
bhāskarīya and a direct commentary on the same author's *Laghubhāskarīya*[65].

The following excerpt has been extracted and translated from his own commentary
to verse 13 of the mathematical chapter of the *Āryabhaṭīya*[66]:

---

[61][Shukla 1976, p.71]
[62][Shukla 1976; p.79]
[63]See [CESS, Volume IV; pp. 187-192]
[64]The first edition of the *Āryabhaṭīya* was published with his commentary: [Kern 1874]
[65][Sastri, 1957]
[66][Kern 1874; p. 32]

"With a *bhrama*, that is, with an instrument (*yantra*) called a *karkaṭa* a circle should be brought about. This is what has been stated:

Having acquired any straight stick (*yāṣti*), having bound it, firmly, with a cord on its upper-part at the throat-spot (*kaṇṭhapradeśa*), having also split ⟨it, vertically⟩ from the lower tip to the throat, ⟨and thus⟩ having made two sticks (*śalākā*), one should make their two tips sharp ones. In this way is produced a *karkaṭa* instrument having an under mouth (or opening *adhomukham*). Having further fixed a stick in the space between the ⟨previous⟩ two sticks one should make a pair of compasses having a revolving opening (*vivṛttāsya*). Having made the *karkaṭa*'s opening equal to the semi-diameter of the desired circle by moving up and down the stick which lies in the intermediate space, having laid the tip of one stick on the central spot of the circle to be brought about, having laid the other tip on the spot on circumference of the circle one should turn the *karkaṭa*. That is the desired circle."

An even more detailed description of the making of a *karkaṭa* can be found in Parameśvara's super commentary to Govindasvāmin's commentary of the *Mahābhāskarīya*. When glossing on verse 1 of the 3rd chapter of this treatise, which describes the circular, flat setting where a gnomon should be placed, Govindasvāmin writes[67]:

> *evaṃ dharātalasya samatvam avagamya*
> *mukhavinyastavartikāṅkuraśobhinā karkaṭena vṛttam ālikhet|*
> Having, in this way, brought evenness to the ground's surface, one should draw a circle with a pair of compasses (*karkaṭa*) beautiful with a sharp stick (*vartikāṅkura*) inlaid at its opening.

Notice that Govindasvāmin uses the compound *vartikāṅkura* which is almost the same expression that we have found difficult to read in the *Āryabhaṭīyabhāṣya*: Bhāskara used the compound *vartyaṅkura*, and once the word *vartikā*, probably as a synonym of *vartī*. Parameśvara glosses the compound used by Govindasvāmin extensively[68]:

"With the word *karkaṭa* an instrument fit for bringing about the outline (*parilekhana*) of a circle is meant. In this case, having acquired any evenly circular stick (*yaṣṭi*), having bound ⟨it⟩ firmly above its middle at the throat spot (*kaṇṭhapradeśa*) with a string (*rajju*), and so on, having furthermore split ⟨it⟩ at its root, one should make it in such a way that below the throat (*ākaṇṭha*) there are two equal sticks (*śalākā*). Afterwards one should make the tip of ⟨each⟩ stick a sharp tip (*tīkṣṇāgra*). This is called a '*karkaṭaka*'.

---

[67] [Sastri ; p.103-104]
[68] *idem.*

The intermediate space between is called "the compasses' mouth (or opening)" (*karkaṭāsya*). Afterwards, having taken another stick (*śalākā*) whose width is bigger than the compasses' sticks (*śalāke*), and whose length is several *aṅgulas*, having cut its two tips, with a knife, one should make a revolving opening (*vivṛttāsya*). In this way, a stick having a mouth (*mukha*) at its two ⟨tips⟩ is called a *vartikāṅkura* (a sharp stick).

Furthermore, having made a revolving-opening-pair of compasses, having placed transversally the sharp stick in its opening, one should place the two sticks of the compasses on the two mouths (*āsya*) of the sharp stick. In this way, having acquired an instrument called a *karkaṭa* adorned with a sharp stick placed at ⟨its⟩ mouth (*mukha*) one should draw a circle with it. Having made the compasses' opening equal to the semi-diameter by moving the sharp stick up and down, having fixed one stick (*śṛṅga*) in the middle of the circle one should turn the other one all around. When made in this way, the desired circle appears.

Or else, with the word *vartikāṅkura* another instrument is meant. When one has placed two iron sticks on the tips of the two sticks of a pair of compasses, that is *vartikāṅkura*. A line is made with that."

We have given a tentative illustration of Parameśvara's two representations of compasses in Figure 32.

Almost 800 years separate Parameśvara's and Bhāskara's commentaries. Most probably compasses underwent technical changes during that lapse of time. Parameśvara has left us a quite precise testimony of what he considered a pair of compasses. Bhāskara, on the other hand, never seems to have been prolific on this subject. We have therefore, rather than letting our imagination run free, echoed Parameśvara's compasses in our translation of Bhāskara's descriptions.

K   BAB.2.14

We will study here the meaning of Āryabhaṭa's verse, attempt to understand the astronomical extension Bhāskara gives to it, and finally will indicate what we can understand of the different gnomons described by Bhāskara.

K.1   Āryabhaṭa's verse

Verse 14 runs as follows:

> *śaṅkoḥ pramāṇavargaṃ chāyavargeṇa saṃyutaṃ kṛtvā|*
> *yat tasya vargamūlaṃ viṣkambhārdhaṃ svavṛttasya||*

Figure 32: A pair of compasses as described by Parameśvara

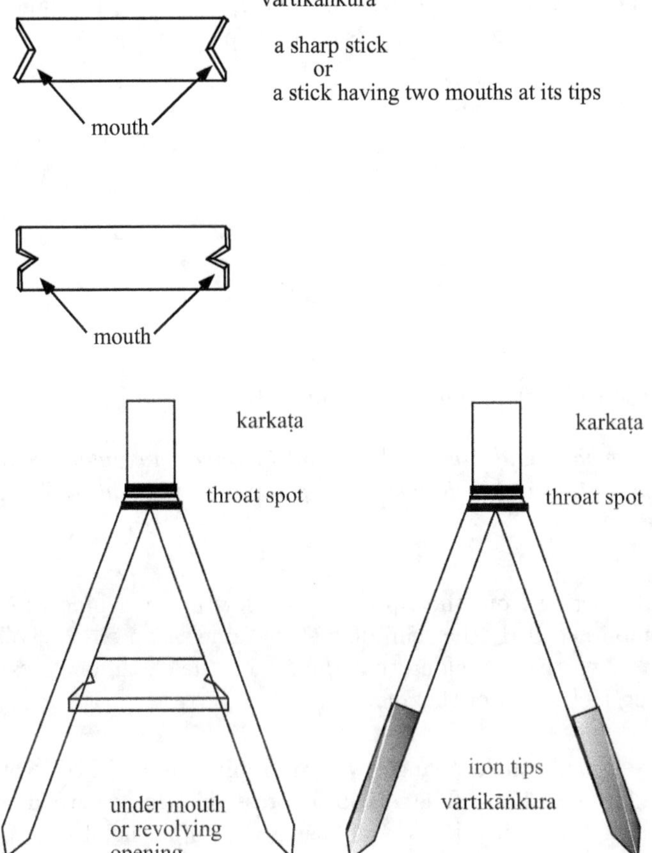

Ab.2.14. Having summed the square of the size of a gnomon and the square of the shadow|
The square root of that ⟨sum⟩ is the semi-diameter of one's own circle‖

The situation described here is the following: a vertical gnomon and the shadow it casts form a right-angle triangle, if we consider the imaginary line that links the tip of the shadow to the top of the gnomon. This imaginary line is called "the semi-diameter of one's own circle" (*viṣkambhārdhaṃ svavṛttasya*). This is illustrated in Figure 33.

Probably, it is by analogy with the celestial sphere – in order to render the ratio, between the gnomon and the position of the midday sun, as we will see in the next section – that the concept of "one's own circle" (*svavṛtta*) is developed. This circle is the one, having the tip of the shadow for center and the distance of the top of

Figure 33: A gnomon, its shadow and the "semi-diameter of one's own circle" OG is the gnomon; OC is the shadow; the circle with C for center and CG for radius is "one's own circle".

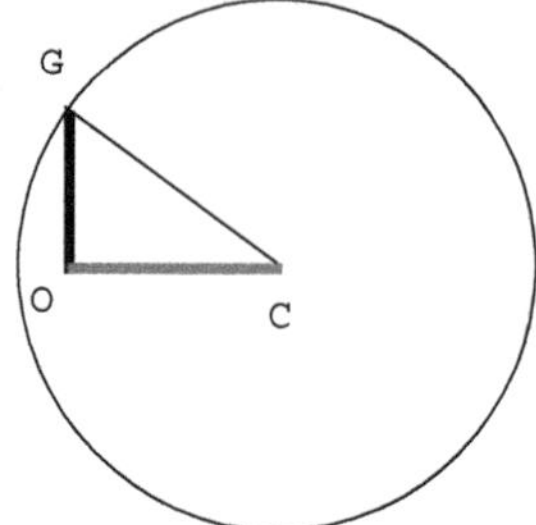

the gnomon to the tip of the shadow for radius. Bhāskara adds:

svavṛttaviṣkambhārdhaṃ nāma chāyāgrād ārabhya śaṅkumastakaprāpi sūtram| tatsūtrānusāreṇa bhūmau dṛṣṭiṃ nidhāya śaṅkumastakāsktaṃ vivasvantaṃ paśyati|

The thread starting from the tip of the shadow and reaching the top of the gnomon is called "the semi-diameter of one's own circle". When one has set down the eye, along that thread, on the earth, one sees the sun adhering to the top of the gnomon.

Because we have a right-angle triangle we can apply the so-called "Pythagoras Theorem", stated in Ab.2.17.ab. The relation expressed in this verse can be written with our modern mathematical knowledge, using the notations of Figure 33:

$$GC = \sqrt{OG^2 + OC^2}.$$

K.2 Understanding Bhāskara's astronomical extension

The astronomical idea behind the use of the gnomon is that the gnomon itself is parallel to the Rsine of the altitude of the sun at mid-day (*Rsinα*), which is thus called by the same name (*śaṅku*). Likewise, the mid-day shadow of the gnomon is parallel to the Rsine of the zenith distance of the mid-day sun (*Rsinz*), both are called *chāyā*[69]. This explains why verticality is an essential feature of the constructed gnomons: the zenith is by definition the point where the line passing through the observer and perpendicular to the horizon, touches the celestial sphere,

---

[69]For a definition of the altitude, the zenith distance, the latitude etc., please see Appendix giving some elements of Hindu astronomy at the end of this volume.

Figure 34: A disproportionate representation of a gnomon and its astronomical interpretation. SuSu' is the orbit of the sun on one particular day. Su is the position of the sun at mid-day. α is the altitude; z is the zenith distance.

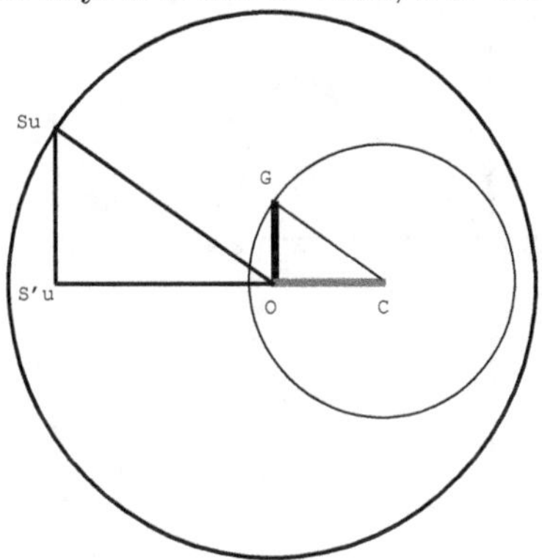

above the observer. It is therefore the verticality of the gnomon that secures that it is parallel to the zenith and therefore to the Rsine of the altitude, and the horizontality of the earth where the shadow is cast, that secures that the mid-day shadow is parallel to the Rsine of zenith distance. In other words, as illustrated in Figure 34, $SuS'uO$ and GOC should form similar triangles.

Knowing the shadow (OC) at mid-day, that is when the sun is on the celestial meridian, and the size of the gnomon (OG) one can compute the Rsine of the altitude or the Rsine of the zenith distance with a Rule of Three.

> *trairāśikaprasiddhyartham– yady asya svavṛttaviṣkambhārdhasya ete śaṅkuc chāye tadā golaviṣkambhārdhasya ke iti śaṅkuc chāye labhyete*

And in this case, the stating of a semi-diameter of one's own circle is ⟨made⟩ in order to establish a Rule of Three: "If for the semi-diameter of one's own circle both the gnomon and the shadow ⟨have been obtained⟩, then for the semi-diameter of the ⟨celestial⟩ sphere, what are the two ⟨quantities obtained⟩?" In that way are obtained the Rsine of altitude (*śaṅku*) and the Rsine of the zenith distance (*chāyā*). Precisely, these two on an equinoctial day are told to be the Rsine of colatitude (*avalambaka*) and the Rsine of the latitude (*akṣajyā*).

In other words, with the same notations as before:

$$\frac{SuS'u}{OG} = \frac{OS'u}{OC} = \frac{OSu}{CG}.$$

In this case, as stated in the Appendix on Some Elements of Indian astronomy,

Figure 35: Gnomon and Celestial sphere

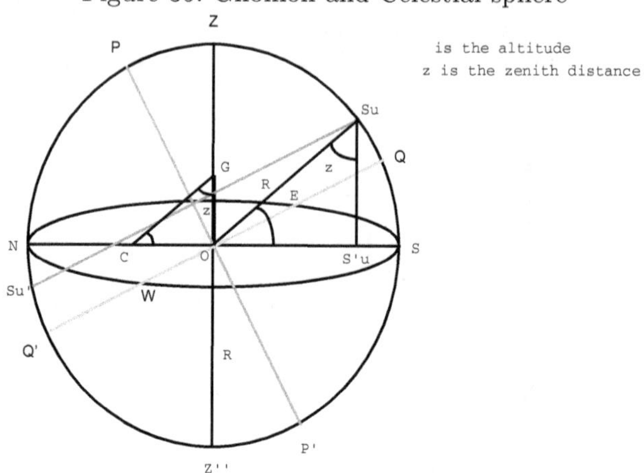

the distance of the observer to the sun is taken to be equal to the radius of the celestial sphere, $R = OSu = 3438$.

Bhāskara adds that other parameters may be computed with this extension of the rule stated in verse 14:

> *chāyayā gaṭikānayane, madhyāhne chāyayā ca sūryānayane*
> *svavṛttaviṣkambhārdhasyāyam eva vidhiḥ|*

When computing the ⟨time in⟩ *ghaṭikās* by means of the shadow and when computing the ⟨altitude of the⟩ sun by means of the mid-day shadow, just that method (*vidhi*) ⟨is used⟩ for the semi-diameter of one's own circle.

We do not know what was the procedure used to compute the time using the shadow of the gnomon, according to Bhāskara or Āryabhaṭa. However, the above ratio can help us understand the sentence that follows:

> *kintu chāyayā ghaṭikānayane śaṅkunā kāryam iti śaṅkur evānīyate|*

However, ⟨this has been told⟩: when computing the ⟨time in⟩ *ghaṭikās* by means of the shadow, ⟨this⟩ should be performed with the Rsine of altitude (*śaṅku*); then just the gnomon (*śaṅku*) is computed.

What should be understood here is that, knowing the Rsine of altitude and the gnomon, then the mid-day shadow can be computed. Using the above ratios, one can reconstruct a probable computation: To compute the mid-day shadow, one uses the following ratio, where OC is the mid-day shadow:

$$\frac{SuS'u}{OG} = \frac{OS'u}{OC} \iff OC = \frac{OG \times OS'u}{SuS'u}.$$

$SuS'u$ is the Rsine of the altitude, and OG the gnomon. $OS'u$ is the Rsine of zenith distance, which does not seem to be requested. But the triangle $OSuS'u$ is right-angled, so that with the "Pythagoras Theorem" we have:

$$OS'u = \sqrt{OSu^2 - SuS'u^2}.$$

OSu is the radius of the celestial sphere, which is a known constant $R = 3438$. So that finally:

$$OC = \frac{OG \times (\sqrt{OSu^2 - SuS'u^2})}{SuS'u}.$$

In the same way, Bhāskara adds:

> *samamaṇḍalacchāyayā sūryānayane sa eva| madhyāhnacchāyayā*
> *sūryānayane natajyayā prayojanam iti chāyaiva ānīyate|*

> When computing the ⟨zenith distance of the⟩ sun with the shadow of ⟨the sun when it is on⟩ the prime vertical (*samamaṇḍala* i.e. at midday), ⟨it is⟩ just like that; when computing the sun with the midday shadow, the Rsine of the zenith distance (*natajyā*) is needed, in this way (*iti*) the shadow (*chāyā*) is computed.

I do not know what corresponds to the "shadow of the prime-vertical", nor what is the coordinate of the sun that was derived from it. But concerning the Rsine of the altitude of the sun, Bhāskara's sentence can be understood as indicating that one just needs to know the Rsine of the zenith distance and the mid-day shadow. We know from the above ratios, where $OS'u$ is the Rsine of the sun's altitude, that

$$\frac{OSu}{CG} = \frac{OS'u}{OC} \iff OS'u = \frac{OC \times OSu}{CG}.$$

OSu is the radius of the celestial sphere, a known constant, and OC the mid-day shadow. CG is the "semi-diameter of one's own-circle" and may not have been requested. But by Ab.2.14 we know that

$$CG = \sqrt{OG^2 + OC^2}.$$

OG is the length of the gnomon and had a standard measure. In Bhāskara's commentary it is always 12 *aṅgulas*. So that in the end we would have

$$OS'u = \frac{OC \times OSu}{\sqrt{OG^2 + OC^2}}.$$

Figure 36: A disproportionate representation of a gnomon on an equinoctial day
The equinoctial mid-day sun is at the crossing point of the celestial equator and
the celestial meridian.

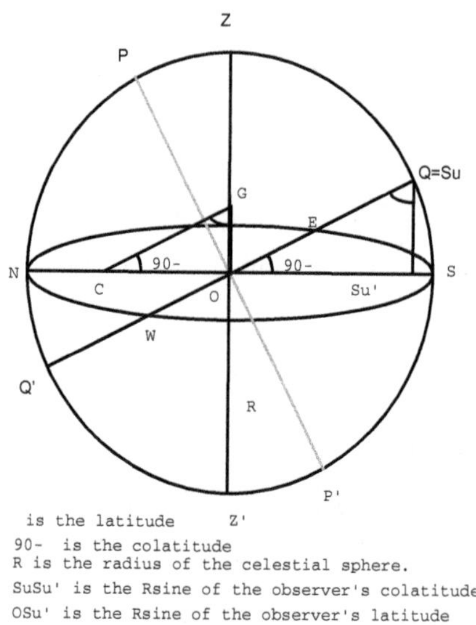

As we have remarked in the Appendix on astronomy, on an equinoctial day, the
sun is on the celestial equator, so that the Rsine of altitude becomes the co-latitude
and the Rsine of the zenith distance, the latitude of the observer and gnomon. This
is illustrated in Figure 36.

Bhāskara states this:

> *tāv eva viṣuvati avalambakākṣajye ity ucyete|*
> Precisely, these two (i.e. the Rsine of the sun's altitude, and the Rsine of
> the sun's zenith distance) on an equinoctial day are told to be the Rsine
> of co-latitude (*avalambaka*) and the Rsine of the latitude (*akṣajyā*)

One can note here that all the values obtained by Bhāskara in the illustrative
examples are approximations.

K.3   Different types of gnomons

Bhāskara describes three types of gnomons in this part of the commentary. These
have been noted and studied by Yukio Ōhashi in [Ōhashi 1994; p.170 sqq]. Our
translation differs at some times from his. This part has remained quite obscure,
and we have just given some tentative representations of such gnomons.

K.3.1 The first gnomon

The first gnomon described by Bhāskara is as follows:

kecit tāvad āhuḥ- dvādaśāṅgulaśaṅkur mūlatribhāge caturaśro, madya-
tribhāge tryaśriḥ, uparitribhāge śūlaākāra iti|
sūkṣmatvād vigrahasya sūkṣmayaikayā koṭiyā chāyāgrasya sulakṣyatvāc
cheṣāiś ca dursampādatvād iti

First, some say: 'A gnomon of twelve *aṅgulas* has four edges on ⟨its⟩
lower third, has three edges on ⟨its⟩ middle third and has ⟨the form
of⟩ a spear on its upper third. Because ⟨the top of the gnomon⟩ has a
sharp shape and because it is easy to characterize a shadow by means of
one sharp upright side and because it is difficult to acquire by all other
⟨means, this is a good gnomon⟩.

From such a description, we do not know what indeed was the shape of the gnomon:
for we do not know how the respective cube, triangular pyramid and the spear
were arranged according to one another. A hypothetical reconstruction is given in
Figure 37.

We do not know how the different shapes (the cube, the pyramid and the spear)
were arranged in respect to one another. Maybe the center of gravity of each object
was on the same line, in which case the pyramid and the cube would have been at
the center of the cube. Here we have assumed that they were all disposed along one
vertical edge of the gnomon, which would therefore be the sharpest. The reasons
why Bhāskara discards such a gnomon, namely that its verticality is difficult to
ascertain, may suggest that indeed, the shape we propose here, is not correct.

Yukio Ōhashi understands the four edged solid to be a prism[70].

K.3.2 The second gnomon

The second gnomon is described as follows:

Apara āhuḥ caturaśraś caturdiśam avalambakasādhanasambhavāt
koṭidvayena chāyāgrahaṇād abhīṣṭakoṭyāṃ dikgrahaṇasiddhir iti|

Others say: " ⟨It should⟩ have four edges because it is possible to bring
about and secure with a plumb-line (*avalambaka*) four directions[71], and,
the knowledge of the direction ⟨of the sun⟩ is established, in the direction
of any desired upright side, from the knowledge of the shadow, with two
⟨opposite⟩ upright sides".

---

[70][Ōhashi 1994; p.171]
[71]Reading *caturdiśām* rather than the *caturdiśam* of the printed edition.

Figure 37: The first gnomon described by Bhāskara

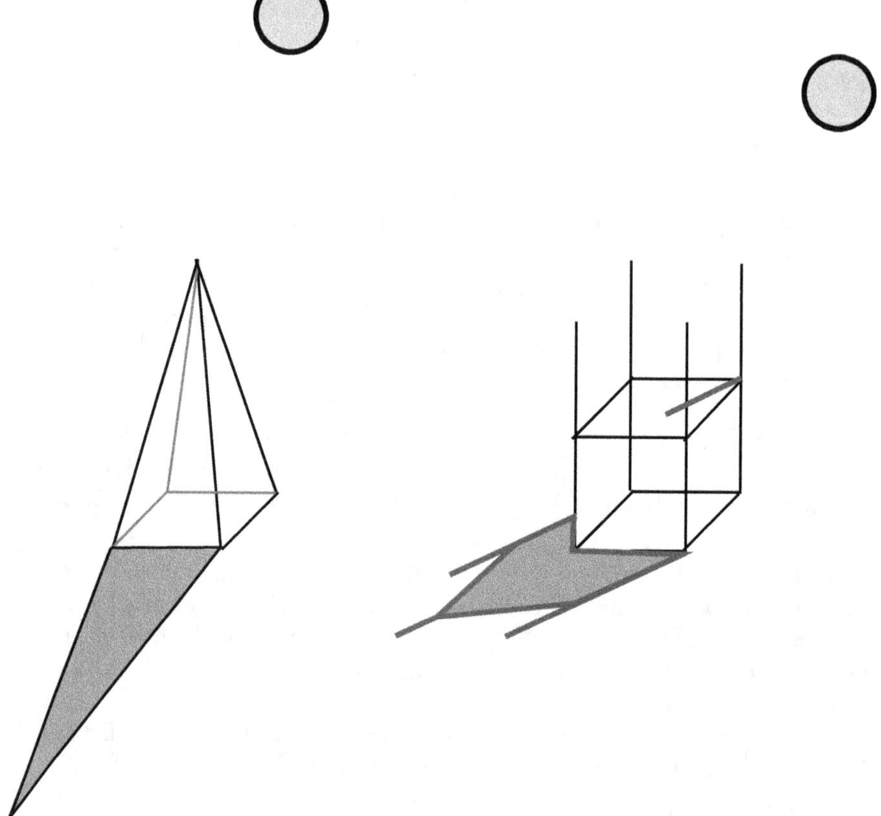

The problem we have in understanding this gnomon, is that we do not know exactly what the "directions" (*diś*), Bhāskara writes about, consist of. Bhāskara rejects this gnomon on the basis that it is difficult to construct, and then adds:

tathāpi pratikṣaṇaṃ sūryasyābhimukhasthāpanāt punaḥ punaḥ śaṅkor mukhacālanaṃ kartavyam| tathā cātisūkṣmadṛśas tāvat (tāvat ābhīṣta) abhīṣṭacchāyātikrāntā syād iti doṣas, etasmāt parityājyo 'yam api śaṅkuḥ| anena eva sarvatra śaṅkavaḥ prayuktāḥ|

Then also because ⟨it should⟩ stand at every moment facing the sun, constantly the face of the gnomon should be made to move. But since, then, for ⟨that gnomon⟩ which at first seems exceedingly precise, the desired shadow will be slightly exceeding, ⟨there is⟩ a draw-back. There-

Figure 38: A hypothetical reconstruction of the second gnomon described by
Bhāskara

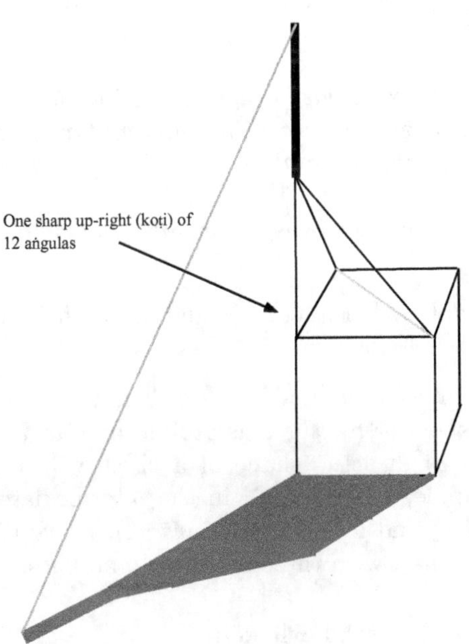

fore, this gnomon also should be set aside. Gnomons are used everywhere
with this very ⟨form⟩.

We have made a hypothetical reconstruction of this gnomon, with its shadow
"'facing the sun" in Figure 38. With such a reconstruction, the reference to the
shadows of two opposite directions makes sense: one appears on the plane in the
middle of the gnomon, the other, parallel to it, and to the two others on the
ground. Why and how the shadow was considered to be "exceeding", I do not
know.

Yukio Ōhashi gives a very different understanding of this gnomon[72]:

> A right prism [whose four sides are] directed towards the four directions.
> For ascertaining the verticality, the shadow of two uprights are made
> coincided (sic), and the direction [of the sun] is ascertained to be in the
> direction of this desired up-right.

---

[72][Ōhashi 1994; p.171]

As he does not give any illustration of such a gnomon, we do not understand
how the prism is oriented "toward the four directions", nor what are the uprights
considered and how they are made to coincide with each other.

K.3.3   The gnomon of Āryabhaṭa's followers

Bhāskara gives the following description of a gnomon according to the "followers
of Āryabhaṭa (*Āryabhaṭīya*):

> *āryabhaṭīyāḥ svamatam abhininiṣṭhāpayiṣavo vyāvarṇayanti|*
> *tad yathā– praśastadārūmayo hy asuṣiro rājigranthivraṇavarjito bhra-*
> *masiddho mūlamadhyāgrāntarālatulyavṛtto nālpavyāso nālpaāyāmaś ca*
> *praśastaḥ|*
> *tribhiś caturbhir vā avalambakair asya ṛjusthitiḥ sādhayitavyā|*

The followers of Āryabhaṭa, wishing to ground firmly their own thoughts,
describe ⟨a gnomon⟩ as follows:

The best ⟨gnomon⟩ indeed is made of excellent wood, has no holes,
is without streaks, knots or fractures; is produced (*siddha*) with a pair
of compasses (*bhrama*), has the shape of a circle which is the same
at the base, the middle, the top, and in the intermediate space; has
a big diameter (*vyāsa*) and a big length (*ayāma*). Its vertical position
(*ṛjusthiti*) is to be secured with three or four plumb lines.

Thus we understand that it is a solid cylinder.

Bhāskara explains then a method to secure verticality:

> *śaṅkum mucce pradeśe niścalaṃ nidhāya avalambakena*
> *śaṅkumūlamastakayor madhye vijñāya tadagrasaktaṃ*
> *prasāryobhayapārśve ca lekhe kūryad| etad ubhayapārśvamadyalekhe,*
> *tataḥ punar api karkaṭakena lohena mūlāgramadhyasūtrābhyāṃ mat-*
> *syam utpādya śeṣamadhyalekhāsādhanam|*

When one has placed the gnomon, firmly, on an elevated spot, having
found the two middle points of the gnomon's base and top respectively,
and having extended a thread fixed to its tip, one should make two lines
on each side (*parśva*). These are the two middle lines (*madhyalekha*)
on each pair of sides; then, once again, having produced, with a pair
of iron compasses (*karkaṭa*), a fish from the two middle threads ⟨which
went through⟩ the base and the top, one secures the remaining ⟨two⟩
middle lines.

Bhāskara also adds on the top a stick, so as to make the shadow of the gnomon as precise as possible. We have not quite understood exactly the construction described here with several threads. A hypothetical reconstruction of this gnomon is given in Figure 39.

Overall a thorough study of the different types of gnomons, and of the meaning of this part of Bhāskara's text, is still needed.

L   BAB.2.15

L.1   Understanding the rule

The situation described by Āryabhaṭa's rule is the following: A gnomon (*śaṅku,* DE) casts a shadow (EC), produced by a source of light (AB). This is illustrated in Figure 40.

The geometrical figure formed by the source of light, the ray of light and the tip of the shadow is a right-angle triangle (ABC). The height of the source of light, is referred to as the base (*bhujā*), and the space between the foot of the light and the tip of the shadow is also called the upright side (*koṭi*). The gnomon (DE) is parallel to AB: its tip, D, lies on the hypotenuse of the triangle and its foot (*mūla*), E, lies on the upright side. BE is the distance between the source of light and the gnomon. The rule given in this verse can be written with the above notations as

$$EC = \frac{BE \times DE}{AB - DE}.$$

This relation is interpreted by Bhāskara as a Rule of Three:

> *etatkarma trairāśikam| katham? śaṅkuto 'dhikāyā uparibhujāyā yadi śaṅkubhujāntarālapramāṇaṃ chāyā labhyate tadā śaṅkunā keti chāyā labhyate|*
> This computation is a Rule of Three. How? If from the top of the base which is greater than the gnomon, the size of the space between the gnomon and the base, which is a shadow, has been obtained, then, what is ⟨obtained⟩ with the gnomon? The shadow is obtained.

With the same notations as before, what is stated is that the ratio of AD to $BE(=DF)$ is equal to the ratio of DE to EC. In other words:

$$\frac{EC}{DE} = \frac{BE}{AB - DE}.$$

If AF on the segment AB represents the distance $AB - DE$, this ratio and the relation given in the verse can be understood as resulting from the similarity of the triangles AFD and DEC.

Figure 39: A hypothetical reconstruction of Āryabhaṭa's followers' gnomon.

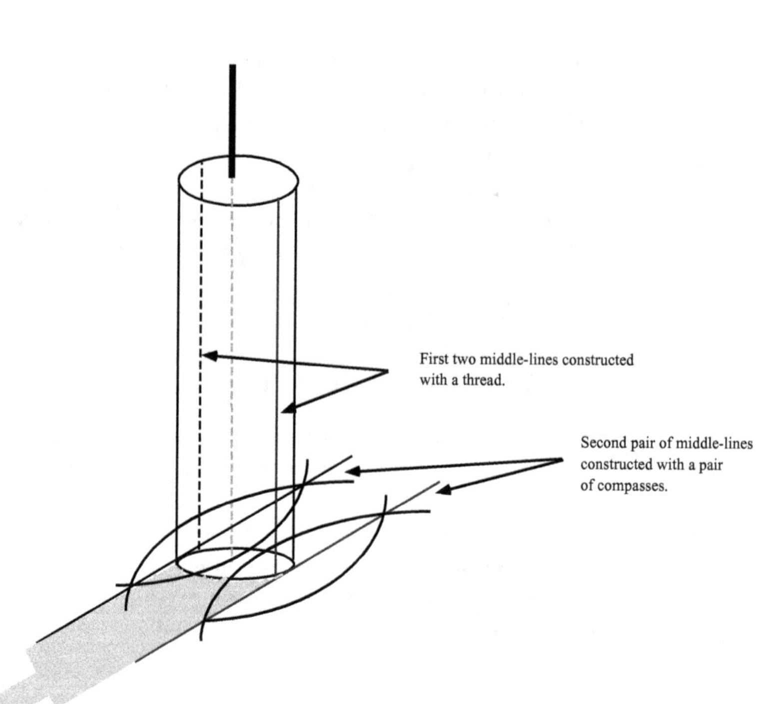

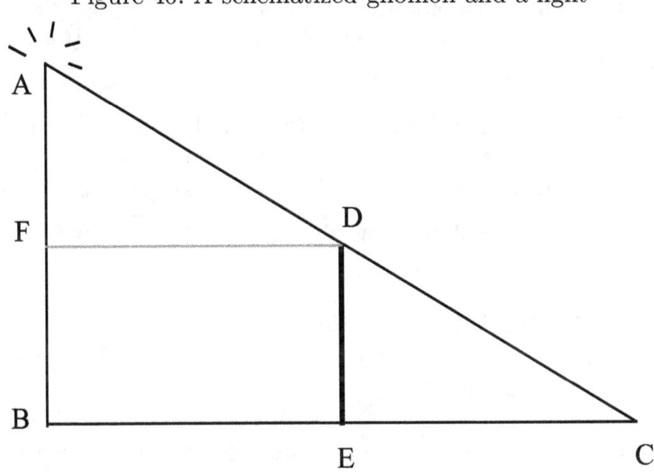

Figure 40: A schematized gnomon and a light

L.2   Procedure

The procedure as it appears in the first versified problem of BAB.2.15 can be summed up by the following steps:

Problem Knowing the height of the light, and the distance separating it from a gnomon of 12 *aṅgulas*, find the length of the shadow cast by the gnomon.

Step1 Multiply the distance between the gnomon and the light by the height of the gnomon ($BE \times DE$).

Step 2 The difference of heights between the light and the gnomon ($AB - DE$) is the divisor of the previous product.

In fact Bhāskara in his commentary, treats all the cases that can appear when a source of light, a gnomon and a shadow are considered. Verse 15 gives a way of finding the length of the shadow knowing the height of the light, of the gnomon, and the distance separating both.

In a second versified problem of BAB.2.15, Bhāskara considers another type of problem: knowing the length of the shadow (EC) cast by a gnomon of 12 *aṅgulas* produced by a source of light of a known height, find the distance (BE) between the light and the gnomon. Bhāskara quotes here the first half of Ab.2.28, which gives a rule to reverse procedures:

> *guṇakārā bhāgaharā bhāgaharāste bhavanti guṇakārāḥ|*
> ⟨In a reversed operation⟩, multipliers become divisors and divisors, multipliers|

In the "procedure" (*karaṇa*) part of the resolution, he explicitly presents the resolution as a way of undoing the computation given in Ab.2.15: one first reverses Step 2 by multiplying by the difference of heights ($AB - DE$), then Step 1 is reversed by dividing by the length of the gnomon (DE).

In the third versified problem given by Bhāskara here, the length of the shadow (EC) cast by a gnomon of 12 *aṅgulas* and the distance separating the gnomon and the light (BE) are known, the height of the light on a pole (AB) is sought.

The resolution here does not use Ab.2.15 at all, but the latter part of the following verse, Ab.2.16[73]:

> *śaṅkuguṇā koṭī sā chāyābhaktā bhujā bhavati*
> That upright side, having the gnomon for multiplier, divided by ⟨its⟩ shadow, becomes the base ‖

Bhāskara reformulates the versified problem in order to show how this rule should be applied:

> *śaṅkubhujāvivarayuktacchāyā koṭir bhavatīti*
> the shadow increased by the space between the base and the gnomon is the upright side.

In other words, as in Ab.2.15, AB is the base, $BC = BE + EC$ is the upright-side.

We can notice that the successive examples solved in this part of the commentary do not function to explain or propound the relation given in Ab.2.15 specifically. They rather seem to examine all the aspects of a given type of problem: with a light on a pole, a gnomon and a shadow, according to the initial values known, different procedures are given in order to deduce the missing values.

M   BAB.2.16.

M.1   Āryabhaṭa's rule

The rule given by Āryabhaṭa in Ab.2.16 involves two computations. This may be understood as follows, according to Bhāskara's interpretation. Let AB be a light disposed on a pole (*yaṣṭi*) whose tip is in A, CD a first gnomon, CH its shadow, EF a second gnomon, whose height is the same as CD, EI its shadow. So the distance between the tips of the two shadows is HI. The distance between the foot of the light and the tip of any of the two shadows (BH and BI), called in Āryabhaṭa's verse "the decrease" (*ūna*), is also referred to as "the earth within the boundary ⟨defined by the foot of the light and the tip of the shadow⟩" (*avasānabhūmi*). This is illustrated in Figure 41.

Figure 41: A source of light with two gnomons

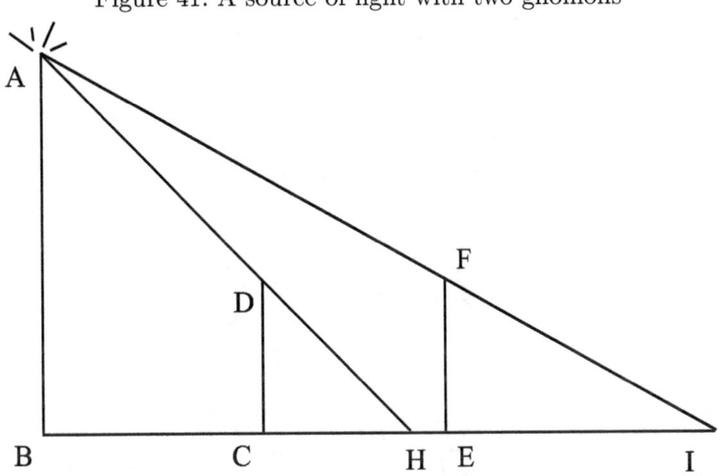

Ab.2.16.ab thus gives a rule to find the distance between the foot of the light and the tip of any shadow, knowing the distance between the tips of the two shadows, the length of one shadow and the difference in length of both shadows.

As in BAB.2.15, the height of the light on a pole is also called the base (*bhujā*). The distance between the foot of the light and the tip of any of the two shadows is also called the upright-side (*koṭī*). Therefore, the presence of two right-angle triangles (*ABH* and *ABI*) is emphasized by underlining.

The generality of the rule, for it applies for either one of the two gnomons, is specified by Bhāskara:

> *tad yadi prathamacchāyayā guṇitaṃ tadā*
> *prathamacchāyāgrayaṣṭipradīpāntarālaṃ bhavati, dvitīyayā chāyayā*
> *yadi tadagrayaṣṭipradīpāntarālam|*
> If that ⟨difference⟩ is multiplied by the first shadow, then ⟨the result of the computation⟩ becomes the space between ⟨the foot of⟩ the light on a pillar and the tip of the first gnomon⟨'s shadow⟩. If that ⟨difference⟩ is multiplied by the second shadow, ⟨then the result becomes⟩ the space between the light on a pillar and that ⟨shadow's⟩ tip.

So this computation can be written as[74]

$$BH = \frac{HI \times CH}{EI - CH},$$

---

[73]Please see the supplement for BAB.2.16. (Volume II, M on the facing page, for an analysis of the use of the rule in this situation

[74]In all cases the examples treated by Bhāskara considers $EI > CH$.

and

$$BI = \frac{HI \times EI}{EI - CH}.$$

These equalities may be understood because of a set of similar triangles: ABH and CDH are similar, therefore

$$\frac{AB}{CD} = \frac{BH}{CH},$$

ABI and EFI are similar so:

$$\frac{AB}{FE} = \frac{BI}{EI}.$$

And since $CD=EF$

$$\frac{BH}{CH} = \frac{BI}{EF} = \frac{BI - BH}{EI - CH}.$$

The second rule given by Āryabhaṭa in the second half of the verse is:

> *śaṅkuguṇā koṭī sā chāyābhaktā bhujā bhavati*‖
>
> That upright side, having the gnomon for multiplier, divided by ⟨its⟩ shadow, becomes the base ‖

With the same notations as before:

$$AB = \frac{BH \times CD}{CH} = \frac{BI \times EF}{EI}.$$

This derives directly from the similarity of triangles and the corresponding ratios as stated above. Ab.2.16.cd gives a rule to find the height of the source of light, knowing the distance between the foot of the light and the tip of any shadow, and the length of that same shadow. Therefore this rule can be applied in the case where only one gnomon is considered: this may explain why it is illustrated in the commentary of verse 15.

Another remark is given by Bhāskara:

> *chāyādvayam api tatkoṭibhyāṃ prasādhyate|*
> The two shadows also are brought about using their two upright sides.

This remark we can understand in other words as stating that if BH and BI are known, both CH and EI can be found.

As this statement is not further developed we can just note here a way of imagining how they were found: the shadows, may have been obtained by reversing the rule given in Ab.2.16.ab. Instead of deriving the upright-side from one of the shadows; one of the shadows is found, knowing one of the upright sides (BH or BI), $EI-CH$ and HI, then:

$$CH = \frac{BH \times (EI - CH)}{HI},$$

$$EI = \frac{BI \times (EI - CH)}{HI}.$$

We can note however that if AB is known, then by similarity of the triangles

$$EI = \frac{EF \times BI}{AB},$$

and

$$CH = \frac{CD \times BH}{AB}.$$

In this case the uprights of the shadows that are referred to would be the heights of the gnomons EF and CD. Now in BAB.2.14 the right-angle triangle formed of a gnomon, its shadow, and the "semi-diameter of one's circle", calls the length of the gnomon the "upright side".

M.2 Astronomical misinterpretations

Bhāskara takes care here to explain how this verse and the previous (Ab.2.15) should not be interpreted astronomically. Many of the arguments he has given remained obscure to us: we will note here what we have understood and what we haven't.

Bhāskara, in order to justify that Ab.2.16 should not be used to find the distance between the sun and the earth, first mentions verse 39 of the fourth Chapter of the *Āryabhaṭīya* (the *golapāda*, chapter on the sphere). This verse computes the length of the "shadow of the earth":

Figure 42: The shadow of the earth

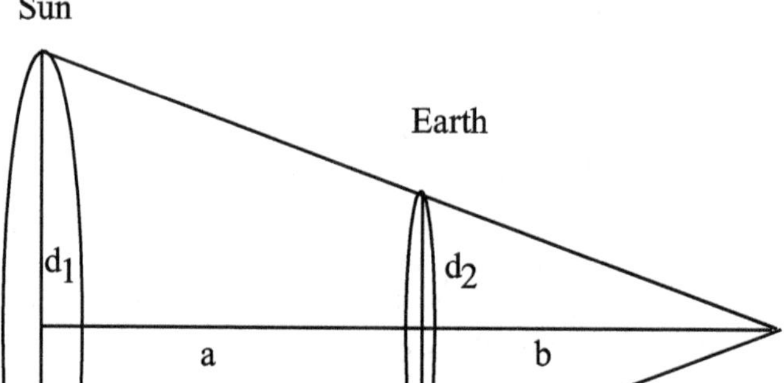

bhūravivivaraṃ vibhajed bhūguṇitaṃ tu ravibhūviśeṣeṇa|
bhūcchāyādīrghatvaṃ labdhaṃ bhūgolaviṣkambhāt
One should divide the distance of the earth to the sun multiplied by the
diameter of the earth by the difference between ⟨the diameters of⟩ the
sun and the earth|
The quotient is the length of the shadow of the earth from ⟨the middle
of⟩ the diameter of the sphere of the earth‖

Let d_1 be the diameter of the sun, d_2 the diameter of the earth, a the distance
between the sun and the earth, b the shadow of the earth. This is illustrated in
Figure 42.

We can write the computation given in the rule as

$$b = \frac{a \times d_2}{d_1 - d_2}.$$

In Ab.1.7 Āryabhaṭa states the diameters of the earth and the sun: $d_2 = 1050$
yojanas (≈ 14360 km[75]) and $d_1 = 4410$ *yojanas* (≈ 60330km[76]). The distance
between the sun and the moon is given by Someśvara in his commentary to Ab.4.39
as being 3360 *yojanas* (≈ 459585 km), consequently the value found for b is 143620

---

[75]Considering that a *yojana* is roughly 13.68 km. For information, we consider today that the
diameter of the earth is 12756 km

[76]Today we consider the diatemeter of the sun to be 14.10^5km.

yojanas ($\approx$ 19666472km[77]).

As we have noted in a footnote in the main translation, we do not know what was Bhāskara's commentary on this verse, but in fact, his commentary on Ab.2.16 looks very much like an explanation of Ab.4.39.

He first gives an interpretation of Ab.4.39 in terms of "gnomons and light on a pole", in which the diameter of the earth, d_2 is considered as the gnomon, the diameter of the sun d_1 is the height of the light, and a is the true distance between the sun and the earth. Because the computation described in Ab.4.39 differs from the one used when reversing Ab.2.15 and described in BAB.2.15, the latter is disqualified for the computation of the distance between the sun and the earth:

> *bhūḥ śaṅkuḥ, raviyojanakarṇaḥ śaṅkubhujāvivaraṃ,*
> *sakalajagadekapradīpo bhagavān Bhāskaraḥ svayam eva*
> *pradīpaucchrāya ity ato vivasvadavanitalāntarālayojanānayanaṃ*
> *na ghaṭate, 'bhūravivivaram' iti siddhānām eva yojanānām upadeśāt|*

> The ⟨diameter of the⟩ earth is a gnomon, the true distance (*karṇa*) in *yojanas* to the sun is the distance between ⟨this⟩ gnomon and the base, the ⟨diameter of the⟩ glorious sun which is the unique light of the whole world, is itself the height of the light. It follows that the computation of the *yojanas* which make the distance between the sun and the surface of the earth is improper, because of the teaching of the *yojanas* that have already been established from the "distance between the earth and the sun". (Ab. 4. 39)

In the paragraph following this interpretation, Bhāskara makes an obscure statement, in which he seems to state that the distinction between the true sun and the mean sun, disqualifies the second part of Ab.2.16 for giving a way of computing the diameter of the sun.

Bhāskara takes additional care to distinguish the case treated in Ab.4.39 both from Ab.2.16, and Ab.2.15. Concerning Ab.2.16 he notes that the configuration as illustrated in Figure 42 does not have two gnomons.

As for Ab.2.15 this results from the sphericity of the earth, as illustrated in Figure 43:the first quarter of verse 6 of the Chapter on the sphere says that the earth is a globe (*vṛtta*)[78].

Bhāskara also speaks about the difficulties that would arise from considering a literally huge gnomon whose shadow would not be properly horizontal because of the natural asperities of the earth. Other obscure parts concern the use of Ab.2.15

---

[77]For information, today we consider that the distance between the earth and the sun is 150.10^6 km.

[78]See [Sharma&Shukla 1976; p.118]

Figure 43: Sphericity of the earth

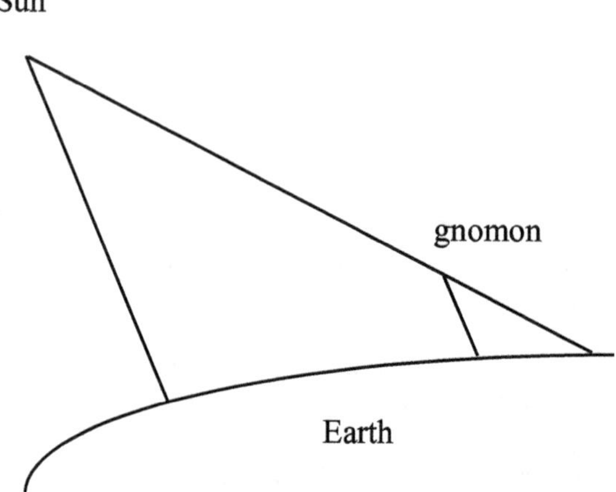

to compute the distance between two cities *Laṅkā* and *Sthāneśvara*. Two other statements are given as improper at the end of the commentary without any explanation. We can note here that *aṅgulas* are much smaller than *yojanas*, considering that a gnomon of 12 *aṅgulas* could give shadows in *yojanas* is nonsensical.

M.3 Ūjjayinī, Laṅkā and Sthaneśvara

From the midday equinoctial shadow of a gnomon, Bhāskara computes here the latitude of Ūjjayinī and Sthāneśvara. From these latitudes he deduces the distance in *yojanas* of these cities to Laṅkā. These three towns are well known to Sanskrit astronomical literature. By definition, Laṅkā is on the intersection of the meridian passing through Ūjjayinī and the equator.

We will, without quoting the text itself, retrace the computation which was made.

M.3.1 Finding the Rsine of Latitude

This operation uses the method described in BAB.2.14 (which we have explained in the Annex of this commentary). Knowing the midday equinoctial shadow of a gnomon (G) of 12 *aṅgulas*, with a Rule of Three using the radius of the celestial sphere (R), 3438 *yojanas*, we can find the Rsine of Latitude $(Rsin\phi)$ for both of these places.

a Ujjayinī The length of the equinoctial midday (C_U) shadow is 5 *aṅgulas*. The length of the "semi-diameter of one's own circle" (R_U) is therefore according to

Ab.2.14:

$$R_U = \sqrt{G^2 + C_U^2} = \sqrt{12^2 + 5^2} = 13.$$

With a Rule of Three the Rsine of Latitude for Ūjjayinī ($Rsin\phi_U$) is deduced:

$$Rsin\phi_U = \frac{C_U \times R}{R_U} = \frac{5 \times 3438}{13} = 1322 + \frac{4}{33}.$$

b Sthāneśvara   With the same type of notations as before (with "S" as superscript), since $C_S = 7$:

$$R_S = \sqrt{G^2 + C_S^2} = \sqrt{12^2 + 7^2} = \sqrt{193}.$$

The result obtained is an irrational number: we do not know how it was approximated. By assuming that $\sqrt{193} \simeq 14$ we have

$$Rsin\phi_S = \frac{C_S \times R}{R_S} = \frac{7 \times 3438}{14} = 1719.$$

M.3.2   Interpolating the Rsinus

In the second half of verse 2 and the first half of verse 3 of the second chapter of the *Laghubhāskarīya*, a method is given to compute an Rsine from a given length, using the table of Rsine differences given in Ab.1.12. This method is translated, explained and discussed in [Shukla 1963; p. 16sqq]. From it we can deduce a method to interpolate the arc whose Rsine we have found, however the results found with this method do not correspond to the values given by Bhāskara. We do not know what method he used. We can note, concerning the latitude of Sthāneśvara, that the value we have found for $Rsin\phi_S$, which depends on the approximate value chosen for $\sqrt{193}$, is 1719. This we know is the Rsine of 30 degrees, as it was stated as such in BAB.2.11. The value given for the latitude of Sthāneśvara is however $30°15'$. The value given for the latitude of Ūjjayinī is $22°37'$.

M.3.3   The distance in between Ūjjayinī, Laṅkā and Sthāneśvara

From what we know of these cities, the distance from Ūjjayinī to Laṅkā would be the measure in *yojanas* of that portion of the terrestrial meridian lying in between Ūjjayinī and the equator: that is a transfer to a value in *yojanas* of the latitude found previously. Using the procedures and values given in this commentary, we know that the diameter of the earth is 1050 *yojanas*. Using BAB.2.10 we can deduce the circumference (*pariṇāha*) of the earth (p) in *yojanas*:

$$p = \frac{62832 \times 1050}{20000} = 3298 + \frac{34}{5}.$$

We can use then a Rule of Three: the ratio of 360 degrees to the circumference of the earth in *yojanas* is the same as the one from the latitude in degrees of Ūjjayinī (22°37′) to the distance (d_U) to Laṅkā in *yojanas*. Converting the values in degrees into minutes, we thus have

$$d_U = \frac{p \times \phi_U}{21600} \simeq 207.4.$$

The value given by Bhāskara is 207 *yojanas*.

Assuming that Sthāneśvara is on the same meridian as Ūjjayinī, the distance of Sthāneśvara to Laṅkā (d_S) is therefore the distance covered by the meridian from Sthaneśvara to the terrestrial equator. With the same reasoning as before,

$$d_S = \frac{p \times \phi_S}{21600} \simeq 274.9.$$

The value given by Bhāskara is 275 *yojanas*.

N   BAB.2.17

N.1   The "Pythagoras Theorem"

Ab.2.17.ab states the so-called "Pythagoras Theorem" in a right-angle triangle. Bhāskara adds in the resolution of example 1:

> *evam adhyardhāśrikṣetre āyatacaturaśrakṣetre vā karṇo yojyaḥ*
> In this way, the diagonal should be considered in a field with an additional half side (*adhyardhāśrikṣetra*) or in a rectangular field.

We do not know what the field "with an additional half side" (*adhyardhāśrikṣetra*) is, as no illustration is given by the commentator. This compound may be connected to the one used in the commentary of Ab.2.17.cd[79]: *ardhatryaśrikṣetra* (a half and a trilateral), which may be referring to the type of field considered within a circle of a right-angle triangle whose upright-side is extended along a diameter, as illustrated in Figure 44.

The additional half side would then be the semi-diameter. Indeed, Bhāskara writes, concerning the trilateral field:

> *ya eva dvitīyo mahāśaraḥ sa eva vaṃśabhaṅgapade*
> *ardhatryaśrikṣetrākāreṇa vyavasthitaḥ|*
>
> That very second large arrow, in the quarter of verse on the breaking of a bamboo, is determined as the shape of semi- ⟨diameter and the side of⟩ a trilateral field (*ardhatryaśrikṣetra*).

---

[79][Shukla 1976; p. 98, line 13]

Figure 44: A right-angled triangle with an additional half-side?

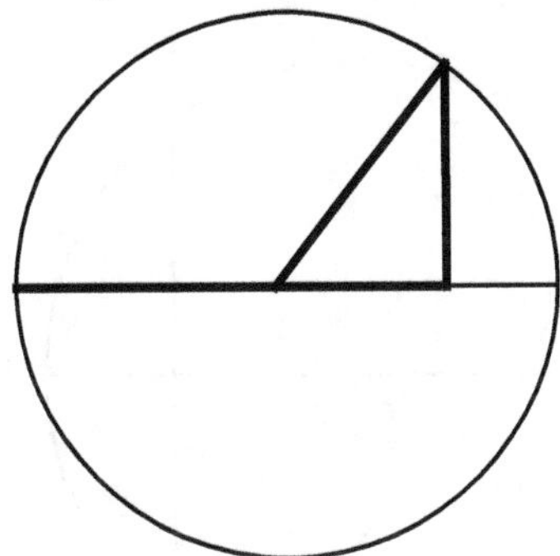

As we will see concerning the interpretation of the second half of verse 17, the "large arrow" is the segment made of the upright side of the right-angle triangle extended by the semi-diameter. The trilateral with an additional half-side referred to here would not be the one illustrated in Figure 45. The latter considers a right-angle triangle having a half-chord for side. The other half of the chord would be the additional half.

Figure 45: A triangle with the other half of the chord

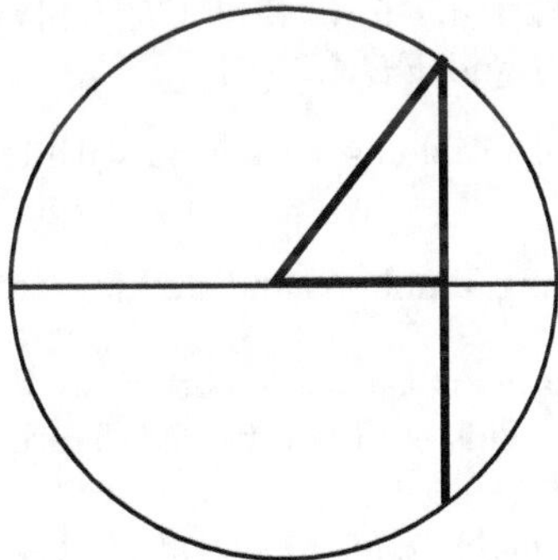

N.2   Two arrows and their half-chord

The second half of Ab.2.17 states the following relation:

vṛtte śarasaṃvargo 'rdhajyāvargaḥ sa khalu dhanuṣoḥ‖
Ab.2.17.cd. In a circle, the product of the arrows that is the square of the half-chord, certainly, for two bow ⟨fields⟩ ‖

Figure 46: Arrows in a circle

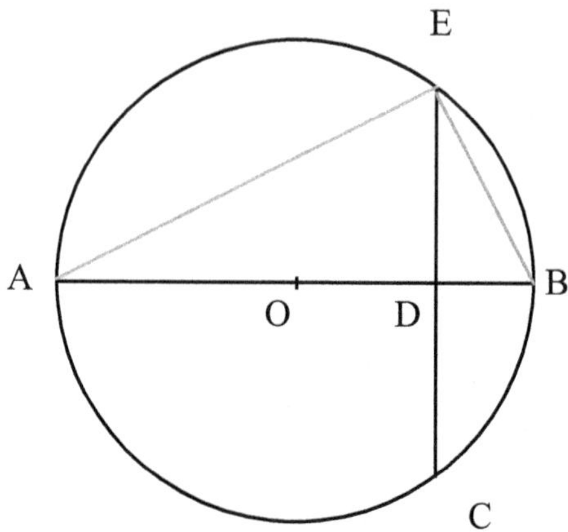

In other words, let a circle of center O have AB a diameter, and CDE a chord. This is illustrated in Figure 46.

We can understand the verse as stating that:

$$DE^2 = AD \times DB.$$

The two "bows" are thus the two arcs formed by $\overset{\frown}{CE}$. This property derives from the similarity of triangles EDB and EDA.

A certain number of traditional problems are solved with this relation.

N.2.1   Rat and Hawks, Breaking Bamboos and Sinking Lotuses

a   The Problems   With the same notation as before, we can state the variety of problems given here as follows. This is illustrated in Figure 47.

Hawk and Rats   A hawk on a height, ED, sees a rat in A whose hole is in D. The rat, seeing the hawk attempts to run back to his hole, but the hawk flying along $E0$ kills the rat at O. Both the distance crossed by the hawk and the distance missing for the rat to reach his hole are sought.

Broken Bamboos   A bamboo of height AD is broken by the wind, it hits the ground at E. The distance between the root of the bamboo and the broken tip is ED. The bamboo is thus formed of two parts, AO and OD, that are sought.

Sinking Lotuses   A lotus is seen above the water, the flower itself being of height DB. It is pushed by the wind for an extent of ED before it sinks. Both the level of the water, OD, and the total size of the lotus, OE are sought.

Figure 47: Hawks and Rats, Broken Bamboos and Sinking Lotuses

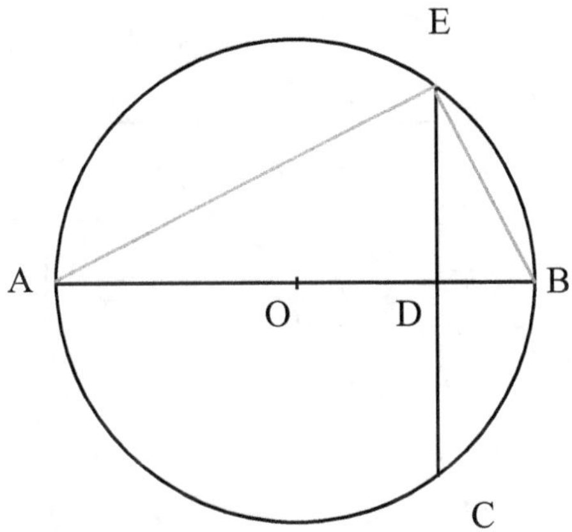

In other words, the general problem treated here is knowing ED and one of the two arrows (AD or DB), both the radius of the circle ($EO = AO$) and OD are to be found.

b procedure Bhāskara states the link between a right-angle triangle and this property of segments within a circle, then relates the computation given in Ab.2.17cd to the "rat and hawk problems"[80].

This computation rests on the fact that

$$DB = \frac{DE^2}{AD}.$$

This is a direct consequence of the computation given in Ab.2.17.cd.

Bhāskara then quotes Ab.2.24. This rule gives a computation called *saṃkramaṇa* which is discussed in the supplement for BAB.2.24.

Precisely, we have

$$EO = \frac{AD + DB}{2},$$

$$OD = \frac{AD - DB}{2}.$$

The first equality computes the radius of the circle $\frac{AD+DB}{2}$, the second one derives from the fact that OD is the radius of the circle decreased by DB.

Bhāskara does not explain in all generality the link Ab.2.17.cd bears with "broken bamboos", but he considers it a simple variation of "rat and hawk" problems[81].

---

[80] See BAB.2.17cd. [Shukla 1976; p. 198, line 3-14]
[81] See the resolution of example 4 [Shukla 1976; p. 100, line 13-15]

Figure 48: Fish and Cranes

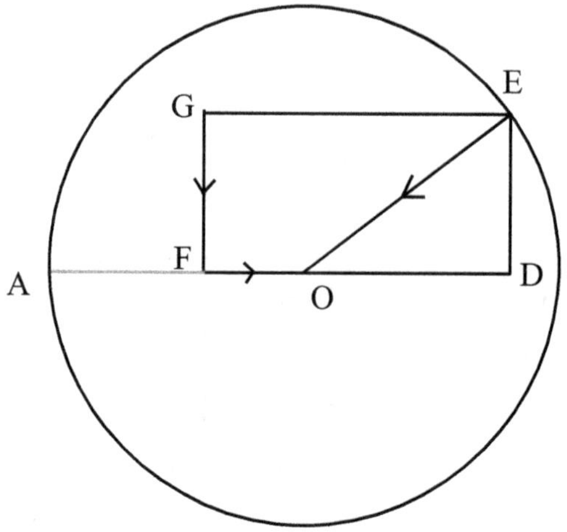

N.2.2   Fish and Cranes

The "fish and crane" problems are slightly different as the setting is in a rectangle.
The problem exposed goes as follows, and is illustrated in Figure 48.

A fish is at one corner of a rectangular tank (E) and a crane at another (G). The
fish crosses the tank diagonally (EO) while the crane walking along the sides of
the tank (GF and FD) catches the fish in O where he is eaten. It is assumed that
the paths of the fish and the crane have the same length ($EO = GF + FO$ in
lengths). It is also assumed that $GF = AF$. Knowing the sides of tank, and the
respective places of the fish and the crane one should find the length of the paths
of the animals (EO), and the distance separating the fish, when it is killed at one
corner of the tank (OD or OF).

Bhāskara relates this situation to the rule given in Ab.2.17.cd as follows:

> *matsyabakoddeśakeṣv apy evam evāyatacaturaśrakṣetrasyaiko bāhur*
> *ardhajyā, bāhudvayaṃ mahāśaraḥ, śeṣaṃ mūṣikoddeśakavat karma|*
> In fish and crane examples, exactly in the same way also, the half-chord
> is one side of a rectangle (ED). The two sides are the greater arrow
> ($GF + FO = AO$), what remains is ⟨as⟩ the method for rat and hawk
> examples.

So that, if we imagine a circle having O for center and EO for radius, as illustrated
in Figure 48, the same procedure as the one for "rats and hawks" will produce
both the radius of the circle $AO = EO$ and OD.

Bhāskara adds at the end of example 8:

pārśvapatite śeṣo dakṣiṇāparakoṇaprāptir matsyasya|

> When the remaining portion of the side is subtracted from the side,
> the remainder is what ⟨was left⟩ for the fish to reach to the south-west
> corner.

In other words, $FO = AO - GF$, since $AF = GF$.

O   BAB.2.18

The rule given by Āryabhaṭa in Ab.2.18 concerns two intersecting circles. Bhāskara
interprets it as concerning an eclipse. The mathematical situation supposed can
be described as follows.

Let two circles intersect in G and F. $ABCDE$ is the straight line, passing through
their respective centers, where AD is the diameter of one circle, BE is the diameter
of the second circle and C the point of intersection of that line with the segment
$[GF]$. The *grāsa* is the segment BD. This is illustrated in Figure 49.

Figure 49: Two intersecting circles

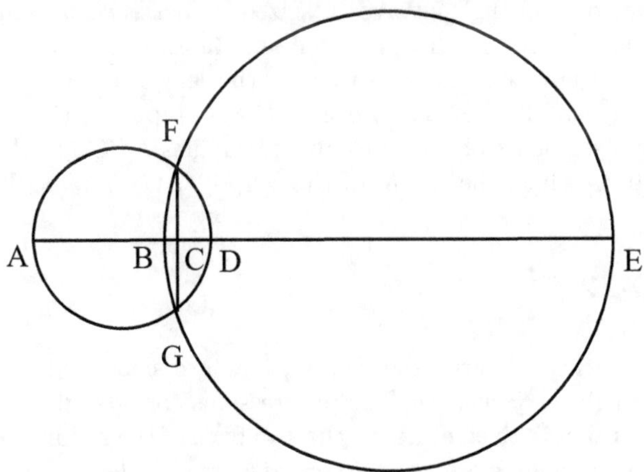

The arrow of the circle AD for the penetrating circle BE is BC. Since $AD - BD =
AB$ and $BE - BD = DE$ it is equal to

$$BC = \frac{AB \times BD}{AB + DE}.$$

And in the same way, for the circle BE:

$$CD = \frac{DE \times BD}{AB + DE}.$$

The ratios linking the segments of the diameters of intersecting circles, with the same notation as before, may be written as follows($BC + CD = BD$):

$$\frac{BC}{AB} = \frac{CD}{DE} = \frac{BD}{AB + DE}.$$

These relations are the key to the rule given in the verse. In this part, Bhāskara does not state them with a Rule of Three as he usually does. He states that the arrow is inversely proportional to the diameter[82]. He also relates the "curving" of the arc and the diameter of the circle. Both explanations are given one after the other. They underline the relation between arrows and diameters, a mathematical interpretation of the word *parasparatah* (one and another) used in the commented verse.

The astronomical context of the verse may be seen through the only versified problem solved in BAB.2.18. It is also given in the following statement by Bhāskara:

> *dve vrtte, grāhyagrāhakamandaladvayam*
> *dve vrtte*, that is two circles , which are the seized and the seizer.

An eclipse (*grahana*) or "seizing" involves a seized planet (*grāhya*) and a seizer (*grāhaka*). In the case of moon eclipses, a demon, Rāhu, is said to swallow the moon. The extent of the eclipse is measured by the length of the *grāsa*, which literally means "a mouthful". Let us note here that computing the "arrows" of the penetration gives segments of the right-angle triangles FBC and FDC (resp. GCB and GCD), from which the extent of the eclipse (FG) may be deduced.

P   BAB.2.19-22

This set of commentaries concerns the rules for progressions and series in the mathematical part of the *Āryabhatīya*. The progressions considered are arithmetical ones. Special attention is given either to the sequence of the natural numbers or to the sequence of their squares or cubes. A *średhī* (series) is defined from the first term of this sequence (*mukha*) and its common difference (*uttara*, lit. increase). The terms of the sequence, and the number of terms of the sequence considered, are both called by Āryabhata *pada*. Bhāskara calls the latter *gaccha*. Different

---

[82]From the above equalities we know that BC (resp. CD) is inversely proportional to DE (resp. AB), which itself is proportional to the semi-diameter.

sums of these terms are considered, which all bear the name *dhana* (value). Because the mathematical computations concerning the sequence always concern the sum of a finite terms of the sequence, *śreḍhī* is translated as "series". If a finite number of terms of the sequence is considered, their sum is called the "whole value" (*sarvadhana*). *madhyadhana* is the "mean value" of the whole sum.

P.1 Ab.2.19

Bhāskara interprets this rule in a very special way. Apparently three rules are given but the first one should not be read literally here. According to Bhāskara's interpretation, the mean value is obtained, as shown below, by omitting the increase by the "previous term". In fact Bhāskara by omitting certain terms reads five rules, in Ab.2.19, that we will expose here.

P.1.1 The mean value

This rule is stated in the first half of Ab.2.19, omitting the computation "increased by the previous ⟨number of terms⟩" (*sa-pūrva*):

> *iṣṭaṃ vyekaṃ dalitaṃ uttaraguṇaṃ samukhaṃ'iti*
> *mdhaydhanānayanārthaṃ sūtram*|
> "The desired ⟨number of terms⟩ decreased by one, halved, having the common difference for multiplier, and increased by the first term", is the rule in order to compute the mean value.

With a modern mathematical notation, let (U_i) be an arithmetical progression of first term (*mukha*) U_1, of common difference (*uttara*) a. Let the "desired number of terms" (*iṣṭa*) be n. By definition the mean value (*madhyadhana*), M, of n terms is

$$M = \frac{\sum_{i=1}^{n} U_i}{n}.$$

M can be computed as follows, according to the rule read in Āryabhaṭa's verse by Bhāskara:

$$M = \left[\frac{(n-1)}{2} \times a\right] + U_1. \tag{6}$$

P.1.2   The value of all terms

By considering the middle part of Ab.2.19, Bhāskara gives a rule to compute the sum of terms in an arithmetical progression:

> '*madhyam iṣṭaguṇitam iṣṭadhanam*' *iti gacchadhanānayanārtham*
> "The mean ⟨value⟩ multiplied by the desired ⟨number of terms⟩ is the value of the desired ⟨number of terms⟩", is ⟨the rule⟩ in order to compute the value of the ⟨desired⟩ number of terms (*gaccha*).

With the same notation as before:

$$\sum_{i=1}^{n} U_i = M \times n. \tag{7}$$

P.1.3   The partial mean value

Two interpretations are given of this rule by Bhāskara: both rest on the ambiguous meaning of *dhana* (value) which can apply to the term of the series (and thus refer to a sum) or to a term of the sequence. A problem occurs because of the discrepancy between Shukla's interpretation of the general rule and the manuscripts, all of which are noted in the main translation.

By omitting the final word evoking the mean value, out of the first half of Ab.2.19, the commentator deduces the following rule:

> '*iṣṭaṃ vyekaṃ dalitaṃ sapūrvam uttaraguṇaṃ samukham*' *ity antyopāntyādidhanānayanārtham*
> "The desired ⟨number of terms⟩ is decreased by one, halved, increased by the previous ⟨number of terms⟩, having the common difference for multiplier, and increased by the first term", is ⟨the rule⟩ in order to compute the value of the last, the penultimate, etc. ⟨terms⟩.

This rule can concern the mean value, M_n, of the sum of n terms (*iṣṭa*: the desired ⟨number of terms⟩) starting with U_{p+1} $-p$ being the previous (*pūrva*) number of terms. By definition

$$M_n = \frac{\sum_{i=p+1}^{p+n} U_i}{n}.$$

This rule would give:

$$M_n = \left[\frac{(n-1)}{2} + p \right] \times a + U_1. \tag{8}$$

If the desired number of terms is 1, this means that the value of one term is computed.

The word *dhana* which literally means "wealth", and which technically in mathematics can mean " value of the terms of a series" or the "value of the terms of the sequence", collects these two meanings here.

In Example 3, which is the only example to illustrate this rule, the former computation is deduced from the first general rule given. The conditions of this example are summed up in the "setting down" (*nyāsa*) part of the solved example:

> *nyāsaḥ– ādiḥ 7, uttara 11, gacchaḥ 25|*
> setting down: the first term is 7, the common difference is 11, the number of terms is 25.

The first part of the resolution seems to describe, in this specific case, how the general rule for the mean value of partial sums may be analyzed to compute the value of specific terms:

> *karaṇam– iṣṭaṃ pañcaviṃśati-(Edition reads viṃśatiḥ) 25, pūraṇaṃ padam*
> *ekam iti ekaṃ rūpaṃ 1, etad eva vyekaṃ śūnyam 0, etad eva sapūrvam*
> *iti śūnyena kṣiptā caturviṃśatiḥ 24, uttaraguṇaṃ 264,*
> *samukhaṃ 271, etad antyadhanam|*
> procedure: the desired ⟨term⟩ is the twenty-fifth term only, and therefore it (the desired number of terms) is one, 25 ; one is unity, 1. Precisely, this decreased by one is zero, 0. Precisely this is "increased by the previous" (here 24), it is increased by zero, and therefore twenty-four, 24, "having the common difference for multiplier", 264, is "increased by the first," 271, this is the ultimate value.

In other words, in the case examined, the number of desired terms, n, is equal to 1. If we substitute 1 for n in the general computation of mean partial sums considered before, then we have:

$$U_{p+1} = M_1 = \left[\frac{(1-1)}{2} + p \right] \times a + U_1 = (0 + p) \times a + U_1.$$

K.S. Shukla gives a different interpretation of this rule. Although all manuscripts read *dalitaṃ* (halved), he omits this word from the rule given here (see p. 105, line 14-15 and note 3), and thus reads:

> '*iṣṭaṃ vyekaṃ sapūrvam uttaraguṇaṃ samukham' ity*
> *antyopāntyādidhanānayanārtham*
> "The desired ⟨number of terms⟩ is decreased by one, increased by the previous ⟨number of terms⟩, having the common difference for multiplier, and increased by the first term", is ⟨the rule⟩ in order to compute the value of the last, the penultimate, etc. ⟨terms⟩.

Thus he understands this rule, as giving only a way of computing the value of each term separately[83]. In other words, if we keep our own notations, he understands the following:

$$U_{p+1} = [(1-1)+p] \times a + U_1.$$

This would explain, step by step, Bhāskara's computation in Example 3, where we have assumed that the halving did not occur, because the numerator was zero.

P.1.4 Partial sum

By quoting the first three quarters of Ab.2.19, omitting the word *madhya* (mean), a rule for the partial sums of the terms of the sequence is derived:

> *'iṣṭaṃ vyekaṃ dalitaṃ sapūrvam uttaraguṇaṃ samukham iṣṭaguṇitam iṣṭadhanam' ity avāntarayatheṣṭapadasaṅkhyānayanārtham|*
> "The desired ⟨number of terms⟩ decreased by one, halved, increased by the previous ⟨number of terms⟩, having the common difference for multiplier, increased by the first, and multiplied by the desired ⟨number of terms⟩ is the value of the desired ⟨number of terms⟩", is a ⟨rule⟩ in order to compute a number of as many terms as desired.

With the same notations as before, if one computes the sum of n terms starting with the term U_{p+1}, then

$$\sum_{i=p+1}^{n+p} U_i = n \times \left[U_1 + a \left(\frac{n-1}{2} + p \right) \right]. \tag{9}$$

P.1.5 Another way of computing the whole value

In the last quarter of the ārya another relation is given:

> *tv athādyantaṃ padārdhahatam||*
> Or else, the first and last ⟨added together⟩ multiplied by half the number of terms ⟨is the value⟩.||

With the same notation as before, if U_1 is the first term and U_n the last, then

$$\sum_{i=1}^{n} U_i = (U_1 + U_n) \times \frac{n}{2}. \tag{10}$$

---

[83] See [Sharma&Shukla 1976; p. 62, formula 3]

P.2   Ab.2.20: The number of terms

The rule given by Āryabhaṭa here, runs as follows:

> *gaccho'ṣṭottaraguṇād dviguṇitādyuttaraviśeṣavargayutāt|*
> *mūlaṃ dviguṇādyūnaṃ svottarabhajitaṃ sarūpārdham||*
> The square-root of the value of the terms (*gaccha*) multiplied by eight
> and by the common difference, increased by the square of the difference
> of twice the first term and the common difference,|
> Decreased by twice the first term, divided by its common difference,
> increased by one and halved.||

Bhāskara gives here a particular interpretation of the word *gaccha* used in the
verse. *Gaccha* is the term used in Bhāskara's commentary for the number of terms
of a series. However concerning its meaning in Ab.2.19, he gives the following gloss:

> *gacchaḥ ity anena [p]adadhanaṃ parigṛhyate|*
> ⟨As for⟩ *gaccha*, the value of the terms (*[p]adadhana*) is understood with
> that ⟨word⟩.

Further in this general commentary, the compound *gacchadhana* is used with the
meaning "the value of the terms"; in this case *gaccha* seems to be a substitute for
pada (a term of a series). This peculiar understanding of *gaccha* is restricted to
this gloss of Ab.2.20. In both cases the compound thus refers to the values of the
terms of the series, or, in other words, to the value of the sum of the terms of a
finite sequence.

Using this particular interpretation, this rule can be understood in the following
way, with a modern mathematical notation:

For a finite arithmetical progression of first term U_1, of common difference a, of
total sum N, the number of terms of the progression is

$$n = \left[\frac{\sqrt{8Na + (2U_1 - a)^2} - 2U_1}{a} + 1 \right] \times \frac{1}{2}.$$

The formulation here is quite surprising. It seems to bear some similarities with
the procedure described in Ab.2.24. These links remain to be investigated.

P.3   Ab.2.21: Progressive sums of natural numbers

Ab.2.20 gives two alternative procedures to obtain the same sum. The first part
of Ab.2.20 runs as follows:

> *ekottarādyupaciter gacchādyekottaratrisaṃvargaḥ|*
> *ṣaḍbhaktaḥ sa citighanas*

> The product of three ⟨quantities⟩ starting with the number of terms
> of the sub-pile whose common difference and first term is one, and
> increasing by one,|
> Divided by six, that is the solid ⟨made⟩ of a pile,

The "sub-pile (*upaciti*) whose common difference and first term is one" corresponds
to the series, (S_i), of the progressive sums of natural numbers[84]: $1, 1 + 2, 1 + 2 +
3, \cdots, 1+2+\cdots+i, \ldots (S_i = 1+2+\cdots+i)$. A finite sequence of this series, starting
with its first term, is considered. Let the number of terms be n. Thus "the product
of three ⟨quantities⟩ starting with the number of terms $(\ldots)$ and increasing by
one corresponds to the product $n(n + 1)(n + 2)$. The "solid ⟨made⟩ of a pile"
(*citighana*), corresponds to the series, (Σ_j) having for terms (S_i) $(\Sigma_j = \sum_{i=1}^{j} S_i)$.
According to this understanding, the computation described above may be noted
as follows:

$$\Sigma_n = \sum_{i=1}^{n} S_i = \frac{n(n + 1)(n + 2)}{6}. \tag{11}$$

The last quarter gives an alternative rule:

> *saikapadaghano vimūlo vā*‖
> Or the cube of the number of terms increased by one, decreased by ⟨its
> cube⟩root, ⟨divided by six produces the same result⟩ ‖

With the same notation as before:

$$\Sigma_n = \sum_{i=1}^{n} S_i = \frac{(n + 1)^3 - (n + 1)}{6}. \tag{12}$$

In his introduction to the chapter on mathematics (*gaṇitapāda*), Bhāskara includes
series (*śreḍhī*) in geometry. A close look at the vocabulary used by Āryabhaṭa and
at the only example of BAB.2.21 may explain how this is understood.

The series (Σ_j) is called by Āryabhaṭa "a solid ⟨made⟩ of a pile" (*citighana*). The
example considers a three-edged pile of objects, of which we have given a tentative
illustration in Figure 50.

We can note here, as it will become clear in the following rules as well, that
the geometrical vocabulary on series is the one used by Āryabhaṭa. Bhāskara
substitutes for it a more arithmetical one, using the term *saṅkalanā* (sum). Thus,
the "sub-pile" (*upaciti*), which corresponds to one layer of the "solid ⟨made⟩ of a
pile", is called *saṅkalanā* by Bhāskara. The "solid ⟨made⟩ of a pile" (*ghanaciti*) is
called *saṅkalanāsaṅkalanā*.

---

[84]As before, the series is constructed considering the sequence which has such first term and
common difference (here the sequence of the natural numbers). The progressive sums of the
terms of the sequence, produces the terms of the series.

Figure 50: "The solid made of a pile"
citighana with 5 layers

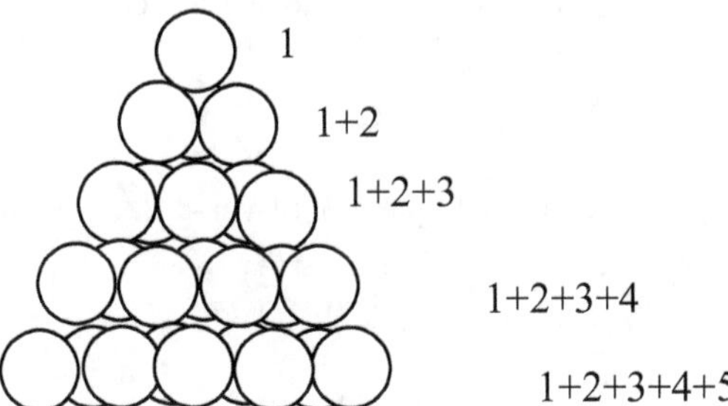

P.4   Ab.2.22: Sum of squares and cubes

Ab.2.22 gives two rules, one for the sum of squares – called the solid ⟨made⟩ of a pile of squares (*vargacitighana*), and one for the sum of cubes – called the solid ⟨made⟩ of a pile of cubes (*ghanacitighana*).

The first rule goes as follows:

> *saikasagacchapadānāṃ kramāt trisaṃvargitasya ṣaṣṭho 'ṃśaḥ|*
> *vargacitighanaḥ sa bhavec*
> One sixth of the product of three ⟨quantities which are⟩, in due order, the number of terms, ⟨that⟩ increased by one, and ⟨that⟩ increased by the ⟨number of⟩ terms|
> That will be the solid ⟨made⟩ of a pile of squares.

Bhāskara gives here a particular gloss of the word usually meaning the term of a progression or series, *pada*:

> *padaṃ gacchas*
> *Pada* is the number of terms.

With this particular interpretation, let n be the number of terms then, with the same notations as before, this rule may be understood as

$$\sum_{i=1}^{n} i^2 = \frac{n(n+1)(n+1+n)}{6} = \frac{n(n+1)(2n+1)}{6}. \tag{13}$$

Bhāskara calls this the sum of squares (*vargasaṅkalanā*).

The second rule is told as follows:

> *citivargo ghanacitighanaś ca‖*
> and the square of a pile is the solid ⟨made⟩ of a pile of cubes‖

This computation uses the rule 10 given in the last quarter of verse 19, which computes the sum of the terms of a finite arithmetical sequence. This is called here "a pile" (*citi*). In this case the sequence considered is that of the natural numbers, zero excluded. Let n be the number of terms considered. According to the rule of verse 19 we know that

$$\sum_{i=1}^{n} i = (n+1) \times \frac{n}{2}.$$

According to the rule given in the last quarter of verse 22, we thus have

$$\sum_{i=1}^{n} i^3 = \left[\sum_{i=1}^{n} i \right]^2.$$

Or in other words:

$$\sum_{i=1}^{n} i^3 = \left[(n+1) \times \frac{n}{2} \right]^2 \tag{14}$$

Bhāskara calls this sum the sum of cubes (*ghanasaṅkalanā*).

Geometrically, the sum of squares, as the diagrams associated to the solved example suggest, seem to be considered as a pile of flat square objects, the smallest having a side of length a unit, the second of length two units etc. In the same way, the sum of cubes seems to be considered as a pile of cubic bricks, the smallest having a side of length one unit etc. These are illustrated in Figure 51.

Q   BAB.2.23-24

Q.1   BAB.2.23: Knowing the product from the sum of the squares and the square of the sum

With a modern mathematical notation, if a and b are any two numbers, the rule given in Ab.2.23 can be summarized as follows:

$$ab = \frac{(a+b)^2 - (a^2 + b^2)}{2}.$$

We can note that when several products are to be computed, each couple is disposed vertically in a column. That is, if the product of a and b and the product of c and d are sought, the "setting-down" (*nyāsa*) will be:

a   c
b   d   .

Āryabhaṭa's verse seems to be useful when the quantities are not known, but only their sums and squares, a typical algebraical problem. Bhāskara's introduction is therefore surprising: he seems to understand it as if it were an alternative multiplication rule. A way of verifying the multiplication algorithm?

Figure 51: Piles

A pile of flat square objects:

the smallest one's area is 1^2

the second one's area is 2^2

etc.

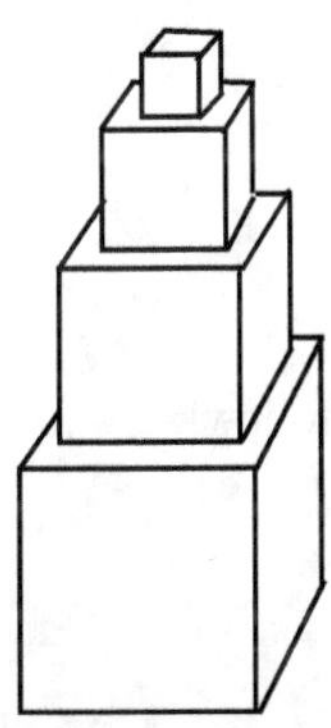

A pile of cubic bricks

The smallest brick's volume is 1^3

The second brick's volume is 2^3

etc.

Q.2 BAB.2.24: Finding two quantities knowing their difference and product

With a modern mathematical notation, this is how the rule given in Ab.2.24 may be understood: let a and b be two quantities $(a > b)$, then

$$a = \frac{\sqrt{(2^2 ab) + (a-b)^2} + (a-b)}{2} ; b = \frac{\sqrt{(2^2 ab) + (a-b)^2} - (a-b)}{2}.$$

The last sentence of the commentary states the commutativity of the multiplication:

> *atra guṇyaguṇakārayor aviśeṣāt guṇakāradvayam ity ucyate|*
> In this case because there is no difference between the multiplicand and
> the multiplier both are called "multipliers".

We can note here that the procedure given in this verse is partially used in other computations. The computation considered involves only the latter half of the verse which involves subtracting or adding to a same quantity a given quantity and halving. It bears the name *saṃkramaṇa*.

R   BAB.2.25

R.1   The rule given by Āryabhaṭa

Ab.2.25 can be formalized as follows. Let m (*mūla*) be a capital; p_1 (*phala*) the interest on m during a unit of time, usually a month, $k_1 = 1$ (*kāla*) ; p_2, the interest on p_1, at the same rate, for a period of time k_2. When $p_1 + p_2$, m, and k_2 are known, in a modern mathematical notation the rule can be understood as

$$p_1 = \frac{\sqrt{mk_2(p_1 + p_2) + (\frac{m}{2})^2} - \frac{m}{2}}{k_2}.$$

This rule derives from a constant ratio:

$$\frac{m}{p_1} = \frac{p_1}{p_2}k_2.$$

We can note that this is algebraically equivalent to the following equation where p_1 is the unknown:

$$k_2p_1^2 + mp_1 - m(p_1 + p_2) = 0.$$

Historians of science have deduced from this that Āryabhaṭa knew how to solve second order equations even though their resolution is not stated as such in the treatise. Second order equations (*vargāvarga*) are quoted by Bhāskara in BAB.1.1., under a list of subjects of mathematics considered in all its generality[85]. However, Bhāskara states a verification of the rule given in Ab.2.25, using a Rule of Five. This rule, therefore, is likely to have been considered by Indian authors as deriving from rules of proportion. The Rule of Five, as described in BAB.2.26-27.ab, and presented in the Annex to this commentary, typically concerns such commercial problems, where k_1 – here always equal to one – may be variable, and where a different value than the initial interest p_1 may be considered as lent at the same rate for a time k_2. The Rule of Five computes a value for p_2:

$$p_2 = \frac{p_1^2 k_2}{mk_1}.$$

A reversed Rule of Five would therefore give a value for p_1, from which the above computation may be found. The Rule of Five, in fact, rests upon the same ratio as the rule given in Ab.2.25, only k_1 may be different from 1:

$$\frac{m}{p_1}k_1 = \frac{p_1}{p_2}k_2.$$

---

[85]See [Keller 2000; I, 2.1] and [Keller forthcoming]

R.2   Procedure followed by Bhāskara in examples

Problem Let m be a capital whose monthly interest p_1 is not known. This interest on the capital is lent elsewhere at the same rate. After k_2 months a certain amount $p_1 + p_2$ is obtained. Both p_1 the initial interest on the capital, and p_2 the interest on the interest are sought.

The tabular "setting down" of such a problem, where the unknown p_1 is noted with a zero, is as follows:

	Interest on the capital	Interest on the Interest
Capital	m	0
Time	1	k_2
Interest	0	$p_1 + p_2$

Step 1 Following the procedure described in Ab.2.25, p_1 is found.

Step 2 The interest on the interest is:

$$p_2 = (p_1 + p_2) - p_1.$$

R.3   Verification with a Rule of Five

Bhāskara at the end of the first solved example of BAB.2.25, describes a verification (*pratyayakaraṇa*). This example states the case where:

$$
\begin{aligned}
m &= & 100 \\
k_2 &= & 4 \\
p_1 + p_2 &= & 6
\end{aligned}
$$

The value found for p_1 is 5.

The verification is stated as follows:

> *pratyayakaraṇaṃ pañcarāśikena– yadi śatasya māsikīvṛddhiḥ pañca tadā*
> *catubhir māsaiḥ śatavṛddheḥ [pañcadhanasya] kā vṛddhir iti|*
> Verification with a Rule of Five: "If the monthly interest on a hundred is five, then what is the interest of the interest [of value-five] on a hundred, in four months?"

In other words, the verification consists in: knowing m, p_1 and k_2, find p_2 and verify that its value, increased by p_1, will give the same value for $p_1 + p_2$ as stated in the problem.

The Rule of Five, as we have stated above, finds the value of p_2. This is how it is

set down:

```
            1    4
set down:  100   5
            5    0
```

So that here the disposition of the Rule of Five follows this pattern:

	Interest on the capital	Interest on the Interest
Time	1	k_2
Capital	m	p_1
Interest	p_1	

However, a regular Rule of Five would be set down as follows:

	Interest on the capital	Interest on the Interest
Capital	m	p_1
Time	1	k_2
Interest	p_1	

So that the two upper rows of the table set-down in BAB.2.25 are inverted. The setting-down of the rule given in Ab.2.25. follows exactly the pattern of a regular Rule of Five.

S   BAB.2.26-27

In this Appendix we will first analyse the procedure given by Āryabhaṭa for the Rule of Three. Afterwards, we will study the rules with several quantities and the Reversed Rule of Three, which are introduced by Bhāskara.

S.1   Rule of Three

Bhāskara treats separately the integral and fractionary cases.

S.1.1   Integers

The Rule of Three is given in verse 26:

> *trairāśikaphalarāśiṃ tam athecchārāśinā hataṃ kṛtvā|*
> *labdhaṃ pramāṇabhajitaṃ tasmād icchāphalam idaṃ syāt||*
> Now, when one has multiplied that fruit quantity in the Rule of Three
> by the desire quantity|
> The quotient of that divided by the measure should be this fruit of the
> desire||

The quantities involved in a Rule of Three and in all the other proportion rules are classified and named according to a typology linked to the kind of problem involved: a measure quantity (*pramāṇarāśi*) produces a fruit quantity (*phalarāśi*),

both are known. A desire quantity (*icchārāśi*) is a new measure quantity whose
fruit, the fruit of the desire (*icchāphala*) also glossed as "the fruit of the desire
quantity" (*icchārāśeḥ phalaṃ*), is sought. The ratio of the first two quantities
is equal to the one of the two others. As we will see in the following subsections,
Rules of Five involve two known measures and desires, Rules of Seven, three known
measures and desires, etc.

According to Bhāskara's commentary, we may describe the Rule of Three as follows:

Problem There is a standard way (called *vāco yukti*) of stating a problem which
involves a Rule of Three, in such a way that the values involved according to
the typology are immediately recognizable:

If by means of a measure (*pramāṇa*, m), a fruit (*phala*, p) has been obtained,
then by means of a desire (*icchā*, i), what is the quantity, called the fruit of
the desire (*icchāphala*, r), obtained?

As studied in [Keller 2000; I.1. 5], we believe that this formulation states
together a mathematical property (equal ratios are involved), serves as a way
of relating a specific problem to the Rule of Three (there are equal ratios in
this problem, which are the following), and prepares the computation of the
Rule of Three, as it underlines the typology of the quantities involved.

Setting-down Bhāskara quotes a verse of unknown origin for the positioning of
the quantities on the working surface:

> *ādyantayos tu sadṛśau vijñeyau sthāpanāsu rāśīnām|*
> *asadṛśarāśir madhye trairāśikasādhanāya budhaiḥ||*
> In order to bring about a Rule of Three the wise should know that
> in the dispositions|
> The two similar (*sadṛśa*) ⟨quantities⟩ are at the beginning and the
> end. The dissimilar quantity (*asadṛśa*) is in the middle.||

The quantities which are "similar" (*sadṛśa*), are those which are similar from
the point of view of the typology of the problem: the measure and the desire,
both of which produce a fruit. The "dissimilar" (*asadṛśa*) one, is the only
known fruit.

The row on which the quantities are written has the following columns:

measure quantity	fruit quantity	desire quantity
m	p	i

Procedure Following the rule given by Āryabhaṭa, the fruit is multiplied by the
desire and divided by the measure:

$$r = \frac{p \times i}{m}.$$

S.1.2 Fractions

As noted before, what is considered a fraction by Bhāskara is a number of the form $a \pm \frac{b}{c}$ noted in the text as:
$$\begin{array}{c} a \\ b \\ c \end{array}$$

Bhāskara glosses the first part of verse 27, to explain how a Rule of Three is carried out with fractions:

> *chedāḥ parasparahatā bhavanti guṇakārabhāgahārāṇām|*
> The denominators are respectively multiplied to the multipliers and the divisor.

This rule would only give the core of the operation to be made when a Rule of Three involves fractions.

Therefore, when fractions are involved, another typology of the quantities entering in a Rule of Three is described. This typology concerns their function within the procedure: a quantity is either a multiplier or a divisor. When multipliers and divisors have denominators, as we will see, their denominators become respectively divisors and multipliers. This property of denominators in the Rule of Three is qualified by Bhāskara as being part of their nature (*dharma*), and the fact of becoming a multiplier or a divisor is stated as a change of condition, by using the verbal root *NĪ-*:

> *... yasmāt taddharmāya chedāḥ parasparaṃ nīyante|*
> ...since according to their nature (*dharma*) denominators are brought
> to one or the other ⟨condition⟩.

We can explain the different steps to be followed according to Bhāskara's commentary as follows:

Problem If the problem is the same as the one stated before, with p, m or i as fractionary quantities:

> If by means of a measure (*pramāṇa*, m), a fruit (*phala*, p) has been obtained, then by means of a desire (*icchā*, i), what is the quantity, called the fruit of the desire (*icchāphala*, r), obtained?

Step 1 All are put into a "same category[86]" (*savarṇita*), which means that the fractionary quantity (i.e a quantity plus or minus a fractional part) is made into a "fraction"[87], with just a numerator and a denominator. So that if p

---

[86] In fact the same word is used in the second half of verse 27 to evoke fractions with a same denominator. See the supplement for BAB.2.27.cd and [Keller 2000; I.2.2]

[87] In [Keller 2000; I.2.2] the status of this intermediary form in respect to fractionary quantities and fractions per se is studied.

was fractionary it becomes of the form $\frac{n_p}{d_p}$, etc. The quantities are "set down" as before, fractions disposed in a column. If all the quantities are fractions the disposition would be as follows:

measure quantity	fruit quantity	desire quantity
n_m	n_p	n_i
d_m	d_p	d_i

Step 2 The computation described in Ab.2.27.ab, as understood by Bhāskara, is carried out. He explains that the denominators of the multipliers (e.g. the denominators of p and i) become divisors and respectively the denominator of the divisor (i.e. the denominator of m) becomes a multiplier.

> *atas teṣāṃ guṇakārabhāgahārāṇāṃ chedāḥ parasparahatāḥ ye*
> *guṇakārachedāḥ bhāgahārahatās te bhāgahārā bhavanti,*
> *bhāgahāracchedāś ca guṇakārahatāḥ guṇakārā bhavanti| (...)*
> *guṇakārāṇāṃ saṃvargo guṇakāra ity arthād avagamyate|*
> Therefore, the denominators are respectively multiplied to those
> multipliers and the divisor; those denominators of the multipli-
> ers which are multiplied to the divisor become divisors and the
> denominators of the divisor multiplied to the multipliers become
> multipliers. (...) Because the meaning is: the product of divisors
> is a divisor; the product of multipliers is a multiplier, ⟨the above
> computation⟩ is understood.

In this particular case, the plural ending of the "denominators of the divisor" may be understood as indicating simply the plurality of unities of the denom- inator (i.e. it is not one). Another interpretation of this plural form as that of a plurality of denominators, makes sense, as we have discussed below, when considering the computation with fractions in rules of proportions involving more than three quantities.

The computation described here involves, probably before the multiplications themselves, a movement of the quantities on the working surface. In Example 2 Bhāskara indicates a movement, by using the verb *Gam-* which means "to go":

> *guṇakārayoś chedā bhāgahāraṃ gatāḥ*
> The two denominators of the multipliers go to the divisor.

So the denominators of the multipliers would move to the column of the divisor and reciprocally, the denominator of the divisor would move to the columns of the multipliers. As there are two columns for the multipliers we do not know where exactly this denominator was placed. We have tentatively represented this movement here:

divisor	Multiplier	
n_m	n_p	n_i
d_p	d_m	
d_i		

No such intermediary disposition, however, is found in the text.

Step 3 As in the procedure described in verse 26, the product of the multipliers is divided by the product of the divisors.

Simplification by common factors were probably commonly used, as among others, the remark quoted below from the resolution of Example 7 shows. We can note that by placing divisors and multipliers in separate but adjoining columns, common factors would appear very clearly.

S.1.3 Conversions

This is a specificity of the problems solved within this commentary: an extra arithmetical computation is required by using different measures of weights and therefore conversions. This may be due to the fact that a quantity, when considered as belonging to a worldly practice (*lokavyavahāra*), should be as much as possible stated as a set of integers rather than with fractional parts. For instance, the result found in the first example has been obtained as a fraction of *palas*, Bhāskara writes:

> *tatra paleṣu bhāgaṃ na prayacchatīti "catuṣkarṣaṃ palam" iti*
> In this case, since parts (*bhāga*) in *palas* are not desired ⟨one should use:⟩ "a *pala* is four *karṣas*"

It is as if the fractional parts themselves occurred because the measuring units were not thin enough. This may have been a common required computation, as even today, measuring units are not uniform throughout the Indian sub-continent. The close link that conversion computations bear with a Rule of Three is underlined in the versified problems concerned with the reversed Rule of Three.

For the sake of simplicity we have not rendered here the computations involving conversions.

S.1.4 Variations

Several examples of problems which involve a Rule of Three are given here. The initial problem is transformed and reformulated in order to make the Rule of Three apparent.

a   Motions In Example 4 the motion of a coiled snake entering its hole is described. The medium speed gives a first ratio between a distance and a time, knowing the size of the snake, the time for the snake to enter the hole is sought.

b Cattle In Example 5, a herd divided in tamed and untamed cattle is considered. Knowing the ratio of the tamed to the total number of heads of one given herd, the ratio to the tamed in another herd is asked. The ratio being thus considered constant from one herd to another.

c Commercial Problems N merchants invest in a capital (*mūla*) each with their own sum: the first merchant invests m_1, the second m_2 etc. The total initial investment is m. The total profit (*lābha*) is known to be l. The respective interests on the initial capital for each merchant is computed with a Rule of Three, the ratio of m to l being the same as the ratio of m_1 to l_1 etc. In example 7, all the initial values are fractionary. They are reduced to a same denominator, Bhāskara then adds:

> *chedaiḥ prayojanaṃ nāstīty aṃśāḥ kevalāḥ*
> There is no use of the denominators, only the numerators ⟨are taken into account⟩.

This makes sense as ratios only are considered.

S.2 Rule of Five and the following

Bhāskara explains at length the case of the Rule of Five, through an example. The Rule of Seven is briefly illustrated. These two cases are sufficient to understand how a rule of $2n + 1$ quantities may be perceived.

These additional rules of proportions may be understood in two separate ways.

1. The first way is only alluded to by Bhāskara in his resolution of Example 8. This is the one we think was usually used in computations, because of its simplicity.

   By generalizing the case presented in Example 8, we can deduce the following situation for a Rule of Five:

   Problem A Rule of Five typically concerns a triple ratio- where two linked measure quantities produce one fruit. This can be expressed as follows: If by means of m_1 and m_2, p is obtained, then by means of i_1 and i_2 what has been obtained?
   Typically this happens in commercial problems, where m_1 is a certain capital invested for m_2 months, p being the interest, i_1 being another capital invested for i_2 months, and r being the interest of this second investment, and is what is sought.

   Setting-down The disposition of a Rule of Five, as seen in the edition[88], has two columns and not three, as in the Rule of Three:

---

[88]One should bear in mind that there certainly is a discrepancy between the edition and the manuscripts on one hand, the manuscripts and Bhāskara's original intentions on the other.

	Known Ratios	Sought Ratios
Capital	m_1	i_1
Time	m_2	i_2
Interest	p	

In one case, in the last row of the second column, where we can suppose that the value sought, r, was placed, a zero (0) is found.[89]

Procedure As in the Rule of Three, where the fruit and the desire are multipliers and the measures are divisors, a rule of Five can be seen as the product of its fruit and desires divided by the product of the measures. In other words:

$$r = \frac{p \times i_1 \times i_2}{m_1 \times m_2}.$$

Accordingly, a greater generalization would consider a rule of $2n + 1$ quantities, typically dealing with the known ratios of n linked measures $m_1, \ldots, m_n$ producing p. Knowing the values of $i_1, i_2, \ldots, i_n$ a value r is sought.

The setting-down would then be of the form:

Known Ratios	Sought Ratios
m_1	i_1
m_2	i_2
.	
.	
.	
m_n	i_n
p	

And the computation:

$$r = \frac{p \times i_1 \times i_2 \times \cdots \times i_n}{m_1 \times m_2 \times \cdots \times m_n}.$$

2. However, Bhāskara insists on the link that rules of proportions have with the Rule of Three. He explains that they all can be understood as a collection of rules of Three:

> *pañcarāśikādīnāṃ trairāśikasaṅghātatvāt| kasmāt pañcarāśy-*
> *ādayas trairāśika saṃhatāḥ ? pañcarāśike trairāisikadvayaṃ*
> *saṃhataṃ, saptarāśike trairāśikatrayaṃ, navarāśike*
> *trairāśikacatuṣṭayam ity ādi*
>
> Because the Rule of Five, etc. is a collection of rules of Three.
>
> ⟨Question⟩
> How are ⟨these rules of Three⟩ to be known?

---

[89]This is presented as such in the edition. We do not know if such a fact was common to all manuscripts. For a discussion, please see [Keller 2000; I.2.2]

> In a Rule of Five, two rules of Three are collected, in a Rule of
> Seven, three rules of Three are collected, in a Rule of Nine, four
> rules of Three are collected, and so forth.

As we have emphasized elsewhere[90], this involves taking each ratio separately,
and considering for each rule the triple first formed of (m_1, p, i_1), and, if p_2
is what is found with a first Rule of Three, then the triple (m_2, p_2, i_2), is
considered and so forth. Thus considering n rules of Three in a rule of $2n+1$
quantities. In the above mentioned paper we have analyzed the way the
disposition itself, by its systematicity, also conveys this somewhat automatic
generation of a new rule, by adding a row.

In the light of these procedures, the discussion Bhāskara carries on the computations with fractions makes sense with several denominators for divisors, since a
rule of $2n + 1$ quantities probably has n divisors. That the computation follows
also the same movement of quantities on the working surface is suggested by the
following sentence at the end of the resolution of example 8 (illustrating a Rule of
Five):

> *chedā api pūrvavad guṇakārabhāgahārāṇāṃ parasparaṃ gacchanti|*
> Furthermore, denominators, as before, go respectively to the divisors
> and multipliers.

S.3   The Reversed Rule of Three

If a Rule of Three may be expressed as

$$r = \frac{p \times i}{m},$$

then the reversed Rule of Three is

$$r = \frac{p \times m}{i}.$$

These ratios typically concern shifting measuring units. We could sum up the
typical problem in this way:

> p of a certain thing has been obtained when measured by a given unit
> u_1 which measures m times another unit, u_2. When u_1 measures i times
> u_2, what is the measure r of the same certain thing, in terms of the unit
> u_1?

---

[90][Keller 1995]

Figure 52: True and mean positions of a planet

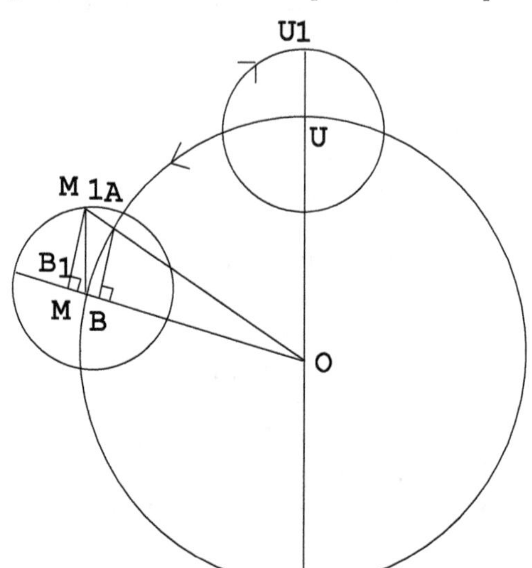

A same quantity is measured with different ratios of conversions, but the unit of measure u_2 is constant, thus that quantity when measured by u_2 amounts to

$$m \times p = i \times r,$$

from which the reversed Rule of Three ratio derives.

An astronomical application of the reversed Rule of Three is given in this part of the commentary. The astronomical parameters it refers to are briefly exposed in the Appendix 3. Bhāskara starts by stating the ratio between two different corrections of the arc distance between the mean position of a planet at a given time (M) and its mean apogee (U):

> *svasiddhānte yadi vyāsārdhamaṇḍale bhujāphalam idaṃ labhyate tadā*
> *tatkālotpannakarṇaviṣkambhārdhamaṇḍale kimiti*
> And in this *Siddhānta* ⟨such a problem requires a Rule of Three⟩ "when this *bhujāphala* has been obtained in the great circle (*vyāsārdhamaṇḍala*), then how much ⟨is it⟩ in the circle whose semi-diameter is the hypotenuse produced at that time?"

Let M_1 be an approximation of the true position of G when its mean position is in M. This is illustrated in Figure 52.

Here Bhāskara does not consider the epicycle defined by MM_1, but the circle having for radius OM_1: *tatkālotpannakarṇaviṣkambhārdhamaṇḍala* (the circle which has for semi-diameter the diagonal produced at that time).

Let A be the point of OM_1 that intersects with the mean orbit of G. Let B be
a point of (MO) such that AB is perpendicular to (MO). Let B_1 be a point
of (MO) such that M_1B_1 is perpendicular to (MO). Both AB and M_1B_1 are
called the *bhujāphala* (the correction to the *bhujā*). OA is the radius of the orbit
(*vyāsārdha*) and OM_1 is called the hypotenuse (*karṇa*).

The ratio given here can be written, with these notations, as

$$\frac{AB}{OA} = \frac{B_1M_1}{OM_1}.$$

Therefore we have:

$$AB = \frac{B_1M_1 \times OA}{OM_1}.$$

Now, Bhāskara remarks that AB is inversely proportional to OM_1:

> *tatra mahati karṇapramāṇo 'lpī yasyo [bhujāphalakalā] bhavanti,*
> *alpakarṇe bahuvya iti*
> In this case when the size of the hypotenuse is great, [the minutes of
> the *bhujāphala*] become smaller, and when the hypotenuse is small, ⟨the
> minutes of the *bhujāphala*⟩ increase.

Now since this portion is stated as he glosses the reversed Rule of Three, we
can understand that Bhāskara, with this relation between the *bhujāphala* and the
hypotenuse, draws a relation from which another analysis of the problem, as a
reversed Rule of Three, could appear: knowing the *bhujāphala* with the radius of
the orbit, we would try to obtain the same segment with the hypotenuse. However
this analysis seems queer as this would suppose that M_1B_1 and AB, which both
bear the same name, are the same segments. This is evidently not the case. The
first *bhujāphala* obtained would be M_1B_1, which seems to derive from the epicycle
and not directly from the radius of the orbit.

Another hypothesis, more convincing maybe, would be to consider that this portion of the text has been displaced in an original scribal error due to the common
ancestor of all manuscripts, and does comment simply on a Rule of Three.

T BAB.2.28

We will discuss here the astronomical computation described by Bhāskara in this part of the commentary.

T.1 Notations and references

Some elements of Hindu astronomy have been given in the Appendix bearing this name. For an understanding of the computation whose steps only are described here, please see [Ōhashi 1994; p. 191-193], [Shukla 1963; p.47-48], [Shukla 1960; p.75-76].

Let the given time in *prāṇas* be t; the ascenscional difference c; the sun's altitude α and the latitude of the observer ϕ, the earthsine (*kṣitījyā*) k, the day radius a.

T.2 Computing the time with the Rsine of the sun's altitude

We can distinguish several steps in the computation given in BAB.2.28, that we may formalize in a modern mathematical form as follows:

1. "when computing the ⟨time in⟩ *ghatis* from the Rsine of altitude produced from the Rsine of zenith distance"

    $Rsin\alpha$ is given

2. "the semi-diameter was a divisor and therefore ⟨becomes⟩ a multiplier"

    $Rsin\alpha \times R$

3. "the Rsine of the observer's co-latitude was a multiplier and therefore ⟨becomes⟩ a divisor"

    $$\frac{Rsin\alpha \times R}{Rsin(90-\phi)}$$

4. "In this case, in the northern ⟨hemi-⟩sphere, one had to add the earth sine, and therefore ⟨it⟩ is subtracted; in the southern ⟨hemi-⟩sphere one had to subtract ⟨it⟩ and therefore it is added. "

    $$\left(\frac{Rsin\alpha \times R}{Rsin(90-\phi)} \right) \mp k$$

5. "Then, just because it has the state of being reversed the semi-diameter is a multiplier"

    $$\left[\left(\frac{Rsin\alpha \times R}{Rsin(90-\phi)} \right) \mp k \right] \times R$$

6. "the day radius is a divisor."

$$\frac{\left[\left(\frac{Rsin\alpha \times R}{Rsin(90-\phi)}\right)\mp k\right]\times R}{a}$$

7. "The sine obtained is made into ⟨its⟩ elemental arcs."

$$Rsin^{-1}\left(\frac{\left[\left(\frac{Rsin\alpha \times R}{Rsin(90-\phi)}\right)\mp k\right]\times R}{a}\right)$$

8. "In the northern ⟨hemi-⟩sphere in arcs the *prāṇas* of the ascenscional difference are added, because ⟨they⟩ had the state of being subtracted; in the southern ⟨hemi-⟩sphere they are subtracted, because they had the state of being added, etc."

$$Rsin^{-1}\left(\frac{\left[\left(\frac{Rsin\alpha \times R}{Rsin(90-\phi)}\right)\mp k\right]\times R}{a}\right)\mp c$$

9. t is obtained.

T.3 Which procedure is reversed?

In verse 28 of the *Golapāda*, Āryabhaṭa gives the following rule[91], which computes the Rsine of altitude from the Rsine of the angular distance of the horizon to the day circle at a given time:

1. Find the Rsine of the arc of the day circle from the horizon (up to the point occupied by the heavenly body) at the given time;

2. Multiply that by Rsine of co-latitude (*avalambaka*)

3. and divide by the radius (*viṣkambhārdha*):

4. The result is the Rsine of the altitude (*śaṅku*) (of the heavenly body) at the given time elapsed since sunrise in the forenoon or to elapse before sunset in the afternoon.

This rule is not exactly the one that Bhāskara reverses, since it doesn't start from the time. However the procedure reversed may have been part of Bhāskara's commentary on this verse. We do not have, however, Bhāskara's commentary on this verse, since none of the known manuscripts have preserved his commentary to the *golapāda*. Procedures are found both in the *Laghubhāskarīya* and the *Māhabhāskarīya*.

---

[91]As translated by Sharma&Shukla 1976, p.139- I have added the indentations and the terms in Sanskrit within parentheses.

In verses 7-11 of the third chapter of the *Laghubhāskarīya* the following algorithm is given[92], which from the time derives the Rsine of altitude:

1. The *ghaṭīs* elapsed (since sunrise) and to be elapsed (before sunset), in the first half and the other half of the day (respectively), should be multiplied by 60 and again by 6: then they (i.e those *ghaṭīs*) are reduced to *asus* (that is to minutes of arc or *prāṇas*[93]:

2. (When the sun is) in the northern hemisphere, the *asu* of the sun's ascenscional difference should be subtracted from them and (when the sun is) in the southern hemisphere, they should be added to them.

3. (Then) calculate the Rsine (of the resulting difference or sum) and multiply that by the day-radius.

4. Then dividing that (product) by the radius (*viṣkambhārdha*), operate (on the quotient) with the earth-sine contrarily to that above (i.e. add or subtract the earth-sine according as the sun is in the northern or southern hemisphere).

5. Multiply that (sum or difference) by the Rsine of the co-latitude (*lambaka*) and divide by the radius: the result is the Rsine of the sun's altitude (*śaṅku*).

6. The square root of the difference between the squares of that and of the radius is the Rsine of the sun's zenith distance (*chāyā*).

7. That multiplied by twelve and divided by the Rsine of the sun's altitude (*śaṅku*) is the true shadow (of the gnomon).

In verses 18-20 of the third chapter of the *Māhabhāskarīya* the following algorithm is given[94]which from the time derives the Rsine of altitude:

1. Add the (sun's) ascenscional difference derived from the local latitude to or subtract from the *asus* elapsed (since sunrise in the forenoon or to elapse before sunset in the afternoon) according as the sun is in the southern or northern hemisphere.

2. (When the sun is) in the northern hemisphere, the *asu* of the sun's ascenscional difference should be subtracted from them and (when the sun is) in the southern hemisphere, they should be added to them.

3. By the Rsine of that (sum or difference) multiply the day-radius,

---

[92][Shukla 1963; Sanskrit text p.11; English translation p.46]. I have added the subdivisions into different steps of the procedure and the names in Sanskrit of terms which occur also in the paragraph of his commentary on verse 28 of chapter 2.

[93]This is the translation adopted by K.S. Shukla of this term (see [Sharma&Shukla 1976; p. 26]).

[94][Shukla 1960; Sanskrit text p.15; English translation p. 74-75]. I have added the subdivisions into different steps of the procedure and the names in Sanskrit of terms which occur also in the paragraph of his commentary on verse 28 of the *gaṇitapāda* of the *Āryabhaṭīya*.

4. and then divide (the product) by the radius (*viṣkambhārdha*).

5. In the resulting quantity apply the earthsine reversely to the application of the ascensional difference (*cara*) (i.e. subtract the earth sine when the sun is in the southern hemisphere and add the earthsine when the sun is in the northern hemisphere).

6. Then multiply that (i.e the resulting difference or sum) by the Rsine of the co-latitude (*lambaka*) of the local place and then divide (the product) by the radius again. Thus is obtained the Rsine of the sun's altitude (*śaṅku*) for the given time in ghaṭīs.

7. The square root of the difference between the squares of the radius and that (Rsine of the sun's altitude) is known as the (great) shadow.

So that with the same notation as before, we can formalize in a modern mathematical language the computation of the Rsine of the sun's altitude with the time:

1. t is given

2. $t \mp c$

3. $Rsin(t \mp c)$

4. $Rsin(t \mp c) \times a$

5. $\dfrac{Rsin(t \mp c) \times a}{R}$

6. $\left(\dfrac{Rsin(t \mp c) \times a}{R} \right) \pm k$

7. $\left[\left(\dfrac{Rsin(t \mp c) \times a}{R} \right) \pm k \right] \times Rsin(90 - \phi)$

8. $\dfrac{\left[\left(\frac{Rsin(t \mp c) \times a}{R} \right) \pm k \right] \times Rsin(90 - \phi)}{R}$

9. $Rsin\alpha$

The fact that each step of this operation is the reverse of the other is illustrated in Table 10.

Table 10: A reversed astronomical procedure

Step	rule to find the $Rsin\alpha$	Step	Reversed rule
1	t is given	9	t is obtained.
2	$t \mp c$	8	$Rsin^{-1}\left(\dfrac{\left[\left(\frac{Rsin\alpha \times R}{Rsin(90-\phi)}\right)\mp k\right]\times R}{a}\right) \pm c$
3	$Rsin(t \mp c)$	7	$Rsin^{-1}\left(\dfrac{\left[\left(\frac{Rsin\alpha \times R}{Rsin(90-\phi)}\right)\mp k\right]\times R}{a}\right)$
4	$Rsin(t \mp c) \times a$	6	$\dfrac{\left[\left(\frac{Rsin\alpha \times R}{Rsin(90-\phi)}\right)\mp k\right]\times R}{a}$
5	$\dfrac{Rsin(t\mp c)\times a}{R}$	5	$\left[\left(\dfrac{Rsin\alpha \times R}{Rsin(90-\phi)}\right)\mp k\right]\times R$
6	$\left(\dfrac{Rsin(t\mp c)\times a}{R}\right)\pm k$	4	$\left(\dfrac{Rsin\alpha \times R}{Rsin(90-\phi)}\right)\mp k$
7	$\left[\left(\dfrac{Rsin(t\mp c)\times a}{R}\right)\pm k\right]\times Rsin(90-\phi)$	3	$\dfrac{Rsin\alpha \times R}{Rsin(90-\phi)}$
8	$\dfrac{\left[\left(\frac{Rsin(t\mp c)\times a}{R}\right)\pm k\right]\times Rsin(90-\phi)}{R}$	2	$Rsin\alpha \times R$
9	$Rsin\alpha$ is obtained	1	$Rsin\alpha$ is given

U   BAB.2.29

The procedure given in Ab.2.29 may be understood, with a modern mathematical computation, as follows:

Two series are defined here. Both use the terms of a given set of quantities, (x_i). One series is obtained with a method that Bhāskara calls "the decreased by a quantity"-method (*rāśyūnakrama* or *rāśyūnanyāya*). This series is defined as follows:

$$S_1 = x_1 + x_2 + \cdots + x_n - x_1 = \sum_{i=1}^{n} x_i - x_1 = x_2 + \cdots + x_n, \cdots$$

$$S_2 = x_1 + x_2 + \cdots + x_n - x_2 = \sum_{i=1}^{n} x_i - x_2 = x_1 + x_3 + \cdots + x_n, \cdots$$

$$S_j = \sum_{i=1}^{n} x_i - x_j,$$

$$S_n = \sum_{i=1}^{n} x_i - x_n.$$

Each term of this series is known. From these terms, the terms of a second series (X_i) are found. Each term X_i consists of the sum of the terms of the set (x_i). In other words:

$$X_i = \sum_{i=1}^{n} x_n.$$

When the series (S_i) is considered, each term x_i appears in every sum except in the sum S_i, thus it appears $n-1$ times. Therefore when the sum of n terms of the series S_i are considered in due order, to obtain X_i one can just divide by $n-1$. In other words:

$$X_i = \sum_{i=1}^{n} x_n = \frac{\sum_{j=1}^{n} S_j}{n-1}.$$

X_i is called the "value of the terms" (*gacchadhana*) or the "whole value" (*sarvadhana*).

Bhāskara however understands that this verse does not only compute the terms of X_i, but also the value of each term x_i, separately. In other words, from both the term X_i and each S_i, all the terms of the set (x_i) called the "value of a term" (*padadhana*) can be found:

$$x_i = X_i - S_i.$$

We note that Bhāskara indicates that the terms of the series S_i should be stored separately in an "undestroyed disposition" (*aviniṣṭasthāpana*) in order to be used in this operation.

V   BAB.2.30

V.1   General resolution of first order equations

This verse gives a procedure to solve first order equations.

Let x be "the price of a bead" (*gulikāmūlya*); it is the unknown. Let a and b be the number of beads belonging respectively to two persons. These are the coefficients of the unknown. Bhāskara also names them *yāvattāvat* (as much as), in which case the unknown is named "the value of the *yāvattāvat*" (*yāvattāvatpramāṇa*). Let c and d be the additional amount of money respectively belonging to each of the two persons. Ab.2.30 reads as follows:

gulikāntareṇa vibhajed dvayoḥ purūṣayos tu rūpakaviśeṣam|
labdhaṃ gulikāmūlyaṃ yadyarthakṛtaṃ bhavati tulyam||
One should divide the difference of coins[95] ⟨belonging⟩ to two men by
the difference of beads.|
The result is the price of a bead, if what is made into money ⟨for each
man⟩ is equal.

Since it is assumed that "what is made into money ⟨for each man⟩ is equal", with
a modern mathematical notation, the problem considered can be noted:

$$ax + c = bx + d.$$

The setting-down of such an equation (*samakaraṇa*) as seen in examples has the
following pattern:

	beads	coins
Person 1	a	c
Person 2	b	d

Āryabhaṭa's verse, in its usual succinct way, indicates that the unknown is found
by dividing the difference of constants (or coins i.e. c and d) by the difference
of coefficients (or beads, i.e. a and b). Bhāskara, firstly, gives a place for the
respective subtracting operations, in the resolution parts of Examples 3 to 5:
the coefficients (or beads or *yāvattāvats*) are subtracted "above" (*upari*) and the
coefficients "below" (*adhas*).

In the first two examples treated by Bhāskara, $b < a$ and $c < d$. So that the
quotient he computes is

$$x = \frac{d - c}{a - b}.$$

The disposition as described in words (no intermediary step is represented) would
then be as follows:

beads   a-b
coins   d-c

Therefore the dividend is below and the divisor above (in fractions the positioning
is reversed).

In the two following examples treated by Bhāskara, $a < b$ and $d < c$, and according
to the intermediary values found, we understand that the following quotient is
computed:

$$x = \frac{c - d}{b - a}.$$

Obviously, a subtraction is always made by removing the smallest quantity from
a larger one.

---

[95] Even though a *rūpaka* is a particular coin, since Bhāskara glosses it with *dīnāra* and in
examples with *dravya*, he probably understands it here as a coin in general.

V.2 Debts and wealth

In the last example a "debt" (*ṛṇa*) is considered, among the number of coins given. In other words, $-c$ or $-d$ may be considered. It is set down, in the printed edition, with a small circle affixed to it (c°), as when a part is subtracted in a fractionary quantity.

The quality of "debt" and "wealth" seems to be only an attribute of the coins at the beginning of the problem. The compound *ṛṇagata* "the state of being a debt" is used once (p. 127; line 10) to qualify the "negative" coin. However, the results of the computation never bear such a quality. A negative/positive quantity appears as the quality conferred to the number of coins, when these coins counted in the evaluation of the total wealth of a person are subtracted or added. The number itself, however is always positive. So it seems that from this part of Bhāskara's text alone, we cannot consider that negative and positive quantities were used in the meaning that we confer to them now.

It is assumed that the wealth of both people is equal, consequently the quotient obtained in the end can never be "negative". It would be meaningless in terms of debts and wealths. As we have seen, the rules given here only work for certain specific cases of equations. We may assume that problems were devised in order to obtain a meaningful result.

To sum it up, we consider here that the notion of "debt" and "wealth" seems to be restricted to the coins which represent the constants of the equation. We do not see them applied to the "beads" or "*yāvattāvats*" which name the coefficient of the unknowns. Nor is it transferred to the "equal wealth" of both people.

Bhāskara quotes a *prakṛt* verse, which is quite corrupted in the manuscripts used for the edition. This rule concerns debts and wealths (*dhana*). To understand it, one should consider first of all that implicitly, a subtraction is always made from the largest quantity in absolute value. Secondly, that this verse concerns the specific computation described in verse 30, as Bhāskara does not specify that its meaning can be extended to other cases. Finally, we think that the quality of "debt" and "wealth" applies to the quantity, and not to the number. We think that the "signs" of the quantities indicate their status in the procedure, signs explicitly what operations quantities should undergo. It does not affect the number itself. According to their nature, the quantities as "positive numbers" are subtracted or added. The result considered is always a "positive number". In other words, we understand the rule given here as describing, according to a typology of the coins in terms of wealth or debt, the different computations to be carried out.

The rule given in BAB.2.30 may be understood as a succession of four rules. Let us consider c and d two positive integers, with $c < d$.

First rule According to Shukla's Sanskrit interpretation, the first rule given is:

> *śodhyam ṛṇād ṛṇam*
> The debt should be subtracted from the debt.

In this case, for us, the subtraction is made from a negative quantity, and a negative quantity is subtracted from it[96]. Thus the operation considered usually when solving the equation would be

$$(-d) - (-c).$$

The rule quoted would indicate that in fact, what is to be computed in this case is

$$c - d.$$

Let us note here, that the final result of the equation is correct, only if $b - a$, $(b - a > 0)$ is computed.

Second rule According to Shukla's Sanskrit interpretation, the second rule given is:

> *dhanaṃ dhanataḥ*
> The wealth ⟨should be subtracted⟩ from the wealth.

Reasoning as previously, when solving the equation, the operation considered would be

$$d - c.$$

This is the case seen in the first two examples of the commentary. The result given is correct only if $a - b, (a - b > 0)$ is computed.

Third and fourth rule According to Shukla's interpretation the third and fourth rule given is:

> *na dhanato na ṛṇataḥ śodhyam| viparīte śodhanam eva dhanaṃ*
> ⟨a debt⟩ should not be subtracted from a wealth, ⟨a wealth⟩ not from a debt|
> When it is reversed, just the subtraction ⟨becomes⟩ wealth.

In other words, when solving the equation, if a negative quantity is to be subtracted from a positive quantity, usually $d - (-c)$ should be computed. The verse indicates, that in this case, the subtraction should be considered as reversed, i.e. it becomes an addition. Therefore $c + d$ should be computed in fact.

When solving the equation, if a positive quantity is subtracted from a negative quantity, that is if $-d - c$ is what should be computed according to the usual rule, the verse indicates that the subtraction should be reversed, that is $c + d$ should be in fact computed. This is the case illustrated in Example 5. The result obtained is correct only if $b - a, (b - a > 0)$ is computed.

---

[96]We do not have any example where such a computation is carried out.

W BAB.2.31

W.1 Understanding the verse

Verse 31 gives a procedure that may be understood in an abstract way. Bhāskara gives two examples in "worldly computations" (*laukikagaṇita*), however his general interpretation is astronomical. We will discuss here this aspect of his interpretation.

Let there be two planets, planet 1 and planet 2 whose respective longitudes at the time of the computation are λ_1 and λ_2, so that the distance between them is $\lambda_1 - \lambda_2$ (hence $\lambda_1 > \lambda_2$). This value is called "*vilomavivara*" when the two planets are going in opposite directions, "*anulomavivara*" when the planets are going in the same directions. Planet 1 stands in the east; planet 2 in the west. The direct motion is from west to east. Their respective motions (*gati*) g_1 and g_2, correspond to the distance they cross in a finite unit of time.

If they are in opposite motions, then the meeting time, Δt will be

$$\Delta t = \frac{\lambda_1 - \lambda_2}{g_1 + g_2},$$

and if they are going in the same direction,

$$\Delta t = \frac{\lambda_1 - \lambda_2}{\mid g_1 - g_2 \mid}.$$

As explained by the commentator, these results are approximate. Bhāskara distinguishes several cases, and explains the rule as a Rule of Three. He also explains how from the time of meeting, the longitude of the meeting spot is found, approximately.

W.2 Bhāskara's distinctions and explanations

Bhāskara justifies the procedure for each of the cases (i.e. when the planets move in opposite directions and when they move in the same direction). The basic idea is that the variation of the distance separating two planets in a given time is, approximately, a constant ratio. This ratio can therefore give an approximate meeting time or longitude of the meeting. This variation of the distance separating two planets in a day is called the "daily passing" (*āhniko bhogaḥ*, noted Δg).

W.2.1 Planets with opposite movements

According to Bhāskara (p. 130; lines 13-14), when two planets move in opposite directions, their "daily passing" is equal to the sum of their motions during a day ($\Delta g = g_1 + g_2$).

Bhāskara then understands the rule to find the meeting time of G_1 and G_2 as a Rule of Three:

> *tena trairāśikakrīyate– yady anenāhnikena bhogenaiko divaso labhyate, tadā 'nena vilomavivareṇa kim iti|*
> A Rule of Three is performed, with that ⟨daily passing⟩: If one day is obtained with that daily passing, then what is ⟨the time obtained⟩ with that distance of ⟨two bodies in⟩ opposite ⟨motions⟩?

With the same notations as before, the ratios understood here are

$$\Delta g : 1 = \lambda_1 - \lambda_2 : \Delta t,$$

so that

$$\Delta t = \frac{(\lambda_1 - \lambda_2) \times 1}{g_1 + g_2}$$

for a time in days, and

$$\Delta t = \frac{(\lambda_1 - \lambda_2) \times 60}{g_1 + g_2}$$

for a time in *ghaṭikās*, since one day is sixty *ghaṭikās*.

W.2.2 Planets moving in the same direction

When two planets are in moving in the same direction, their "daily passing" is equal to the difference of their motions ($\Delta g =\mid g_1 - g_2 \mid$). Bhāskara states this rather elliptically:

> *yadā punar anulomagatī etau bhavatas tadā bhuktiviśeṣeṇānulomavivarasya bhāgaḥ, yasmād bhuktiviśeṣatulyam āhnikaṃ gatyantaraṃ tayoḥ|*
> Furthermore, when both ⟨planets⟩ are in a direct motion, then, the division of, the distance of ⟨two bodies⟩ with a direct ⟨motion⟩, by the difference of daily motions ⟨is made⟩, because the difference of daily motions is equal to their daily difference of motions.

Once again, a Rule of Three is stated considering in this case a time given in *nāḍīs* or *ghaṭikās*, being two different names for the same measuring unit:

> *tato 'nena gatyantareṇa bhuktiviśeṣeṇa janitena ṣaṣṭir nāḍyā upalabhyante tada anulomavivareṇa kim iti ghaṭikā labhyante|*
> Then, ⟨if⟩ sixty *nāḍī* are obtained with that ⟨daily⟩ difference of motions, produced as the difference of daily motions, then what ⟨is the time produced⟩ with the distance of ⟨two bodies with a⟩ direct motion?

This Rule of Three, would express the ratio:

$$\Delta g : 60 = \lambda_1 - \lambda_2 : \Delta t.$$

So that for a time in *ghaṭikās*:

$$\Delta t = \frac{(\lambda_1 - \lambda_2) \times 60}{\mid g_1 - g_2 \mid}.$$

That this meeting time is an approximation is clearly stated in the first verse that Bhāskara quotes from his own astronomical treatise, the *Mahābhāskarīya*. This verse also explains that because of this approximation, the determination of the longitude of the meeting point requires extra work.

W.3   Finding the longitude of the meeting point

Using the same type of ratios, Bhāskara also gives a rule to find the longitude of the meeting point of two bodies.

W.3.1   Two planets moving in opposite directions

Concerning two bodies with opposite directions, Bhāskara describes the following case:

> *yadaiko grahaḥ purastāt sthito vakrī, [anyaḥ] paścād avasthitaś cāreṇa*
> *gacchati, tayor antarālaliptā vilomavivaram|*
> When one planet, standing in the east, goes in a retrograde ⟨motion⟩ and [the other], existing in the west, goes in an ⟨ordinary⟩ motion; the minutes (*liptās*) of the interval (*antarāla*) ⟨separating them⟩ is "the distance of ⟨two bodies moving in⟩ opposite directions".

The situation described here is illustrated in Figure 53.

Let λ_0 be the meeting spot of G_1 and G_2, and $\Delta\lambda_i = \mid \lambda_0 - \lambda_i \mid$, for i=1,2. Bhāskara states the following Rule of Three, once one has found the meeting time in *ghaṭikās* of G_1 and G_2 (Δt):

> *yady ṣaṣyā ghaṭikābhiḥ grahasphuṭagatir labhyate,*
> *tadā vilomotpannaghaṭikābhiḥ kā bhuktir*
> If the true ⟨daily⟩ motion of a planet is obtained with sixty *ghaṭikās*, then what is the motion ⟨obtained⟩ with the *ghaṭikās* known ⟨as the meeting time of two planets with⟩ opposite ⟨motions⟩?

With the same notation as before, the ratio expressed here would be

Figure 53: Two planets moving in opposite directions, the second having a direct motion

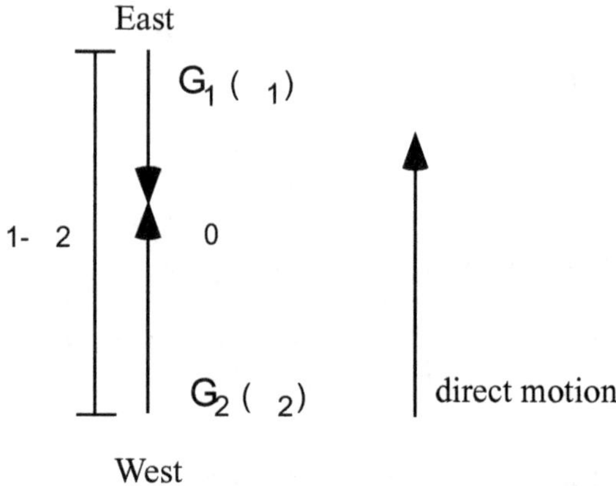

$$60 : g_i = \Delta t : \Delta\lambda_i,$$

so that

$$\Delta\lambda_i = \frac{g_i \times \Delta t}{60}.$$

Once again this ratio may be understood if we consider that numerically g_i corresponds to the distance crossed by planet P_i during a day or 60 *ghaṭikās*.

Bhāskara then states:

> *labdham anulomagatau grahe prakṣipyate vilomagater apanīyate|*
> What is obtained is summed into the ⟨longitude of⟩ the planet with a direct motion, or subtracted from ⟨the planet with⟩ a retrograde motion.

That is, if planet 1 is going in a retrograde motion and planet 2, goes in a direct one, with the same notation as before:

$$\lambda_0 = \lambda_1 - \Delta\lambda_1 = \lambda_2 + \Delta\lambda_2.$$

$\Delta\lambda_i$ represents the correction of longitudes, taking in account the approximate meeting time obtained.

Bhāskara describes a second case of planets moving in opposite directions, illustrated in Figure 54.

Figure 54: Two planets moving in opposite direction; The first planet has a direct motion

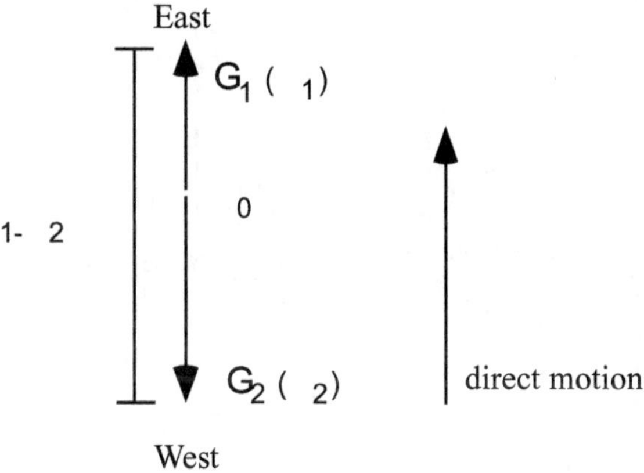

With the same notations as before, and with the same type of reasoning, considering that planet 1 is going in a direct motion, and planet 2 is going in a retrograde motion:

$$\lambda_0 = \lambda_1 - \Delta\lambda_1 = \lambda_2 + \Delta\lambda_2.$$

W.3.2 Two planets moving in the same direction

Bhāskara, in the case of two planets moving in the same direction, distinguishes between one that goes faster than the other. Whether they are in direct or retrograde motion is explicitly stated to be irrelevant:

> *labdhaṃ śīghragatau paścād vyavāsthite ubhayam ubhayatra svaṃ svaṃ*
> *prakṣipyate| śīghragatau puraṣ sthite tad ubhaym ubhayasmād apanīyate|*
> When the ⟨planet with⟩ a faster motion stands westward ; the pair is
> added into the pair, respectively. When the ⟨planet with⟩ a faster motion
> stands eastwards, that pair is subtracted from the pair. In this way the
> past or future meeting times of both are produced.

The situation described in both cases is illustrated in Figure 55.

In this case, the pairs referred to are probably the results obtained for each planet, respectively in the Rule of Three (i.e. $\Delta\lambda_1$ and $\Delta\lambda_2$). We can transcribe the reasoning in a modern mathematical language. If $v_1 < v_2$, then

$$\lambda_0 = \lambda_1 + \Delta\lambda_1 = \lambda_2 + \Delta\lambda_2,$$

Figure 55: Two planets moving in the same direction

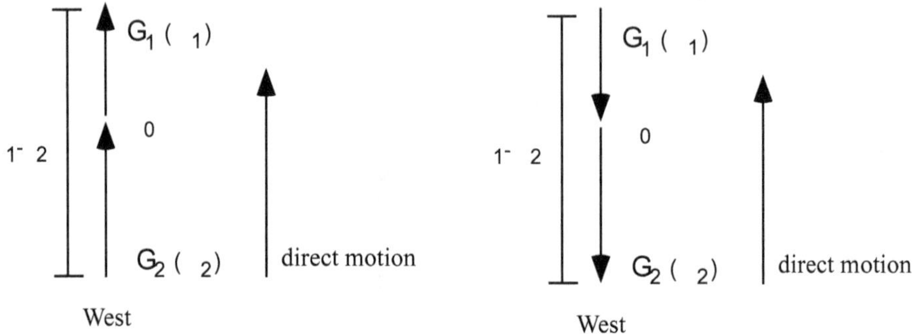

and if $v_1 > v_2$,

$$\lambda_0 = \lambda_1 - \Delta\lambda_1 = \lambda_2 - \Delta\lambda_2.$$

X   BAB.2.32-33: The pulverizer

Bhāskara has two general interpretations of the procedure given in verses 32-33 that describe a "pulverizer computation" (*kuṭṭākāragaṇita*). He reads in these verses a "pulverizer with remainder (*sāgrakuṭṭākāra*)" and a "pulverizer without remainder (*niragrakuṭṭākāra*)". Having explained and illustrated these two different interpretations, he then gives a long list of solved examples which show how one or the other procedure is used in an astronomical context[97].

We will describe and comment on the two different procedures given by Bhāskara, and then we will explain the many astronomical situations in which he applies them. Descriptions, under the label "General comments", will use a symbolical algebraization of the problem.

X.1   Two different problems

The problems that a pulverizer "with remainder" and that a pulverizer "without remainder" solve, are different but nevertheless equivalent.

Indeed, the problem solved by a pulverizer "with remainder" is the following:

What is the natural number N that divided by a leaves R_1 for remainder and divided by b leaves R_2 for remainder?[98]

---

[97] For Āryabhaṭa's and Bhāskara's treatment of the pulverizer, see [Jain 1995; p. 422-447]

[98] Concerning the conditions under which this problem is solvable, please see section X.2.2 of this supplement.

In a modern mathematical language:

$$N = ax + R_1 \quad 0 \le R_1 < a$$
$$N = by + R_2 \quad 0 \le R_2 < b$$

The problem solved by a pulverizer "without remainder" is the following:

What is the integer x that, multiplied by a, increased or decreased by c and divided by b, produces an integer y?

In other words the problem consists of finding two integers (x, y) that verify

$$y = \frac{ax \pm c}{b},$$

where a, b and c are known positive integers. x is called the pulverizer or the multiplier (*guṇaka*), y the quotient (*labdha*).

If we consider the problem solved by a pulverizer with remainder: $R_1 > R_2$, and $R_1 - R_2 = c$, then

$$\begin{cases} N = & ax+ & R_1 \\ N = & by+ & R_2 \end{cases} \Leftrightarrow y = \frac{ax + c}{b}$$

What is called "the divisor of the greater remainder" (a) in the pulverizer with remainder process is called in the pulverizer without remainder "the divisor which is a large number" or "the dividend"; what is called "the divisor of the smaller remainder" in the procedure of the pulverizer with remainder is called here "the divisor"; and what is called the "difference of remainders" ($R_1 - R_2$) is called "the interior of a number".[99]

As we will see, the pulverizer with remainder transforms the problem it solves into a pulverizer without remainder problem. Both procedures, therefore, share common steps. However the two problems and their two procedures are separated in Bhāskara's commentary.

We will now describe the process followed for a pulverizer without remainder.

X.2   Procedure for the pulverizer "with remainder"

We will present here the different steps of this algorithm. We will then expose some of its variations as observed in solved examples, and finally present a mathematical analysis of it.

---

[99] For a brief description of how Bhāskara proceeds to give two different interpretations of the same compound see [Keller 2000; Volume I, I] and in Volume I, Introduction.

X.2.1   General case

Problem

The problem this procedure solves is the following:

What is the natural number N that divided by a leaves R_1 for remainder and divided by b leaves R_2 for remainder?[100]

In a modern mathematical language:

$$N = ax + R_1 \quad 0 \le R_1 < a$$
$$N = by + R_2 \quad 0 \le R_2 < b$$

For $R_1 > R_2$ the "setting-down", in examples, follows this pattern: $\begin{array}{cc} R_2 & R_1 \\ b & a \end{array}$

Step 1

Sanskrit *Ab. 2.32ab. adhikāgrabhāgahāraṃ chindyād ūnāgrabhāgahāreṇa*

English Ab. 2.32ab. One should divide the divisor of the greater remainder by the divisor of the smaller remainder.

General Comments Supposing $R_1 > R_2$, then a is "the divisor of the greater remainder", and b is "the divisor of the smaller remainder"; the following computation is then carried out:

$$\frac{a}{b} = q_1 + \frac{r_1}{b} \quad \Leftrightarrow \quad a = bq_1 + r_1$$

We can note that Bhāskara in examples describes the result as follows: "the remainder is r_1 above, b below". This is probably a way of describing the fractional part that the division produces.

Step 2

Sanskrit *Ab.2.32c. śeṣaparasparabhaktam*

English Ab.2.32c. The mutual division ⟨of the previous divisor⟩ by the remainder ⟨is made continuously.⟩

General comments In other words, the following successive divisions are carried

---

[100]Concerning the conditions under which this problem is solvable, please see the last part of this section of the supplement BAB.2.32-33.

out:

$$\frac{b}{r_1} = q_2 + \frac{r_2}{r_1} \quad \Leftrightarrow \quad b = r_1 q_2 + r_2$$

$$\frac{r_1}{r_2} = q_3 + \frac{r_3}{r_2} \qquad r_1 = r_2 q_3 + r_3$$

$$\frac{r_2}{r_3} = q_4 + \frac{r_4}{r_3} \qquad r_2 = r_3 q_4 + r_4$$

$$\vdots$$

$$\frac{r_{n-2}}{r_{n-1}} = q_n + \frac{r_n}{r_{n-1}} \qquad r_{n-2} = r_{n-1} q_n + r_n$$

No indication is given concerning how to end the process. The "procedure" parts of solved examples suggest that it was stopped when the remainder obtained was considered sufficiently small, i.e. before zero was obtained as remainder. We do not know according to what criteria a quantity was considered to be small enough.

Step 3

Sanskrit *Ab.2.32cd matiguṇam agrāntare kṣiptam*

English Ab.2.32cd ⟨The last remainder⟩ having a clever ⟨quantity⟩ for multiplier is added to the difference of the ⟨initial⟩ remainders ⟨and divided by the last divisor⟩.

General comments As we will see in the next step, Bhāskara indicates how the clever quantity should be placed in regard to the previously computed remainder. The placement presupposed, though not explicitly mentioned, would be:

$$q_2$$
$$q_3$$
$$\vdots$$
$$q_n$$

Bhāskara adds the following gloss which explains under what conditions and how the "clever ⟨quantity⟩" is found[101]:

> *matiguṇaṃ, svabhuddhiguṇam ity arthaḥ|*
> *kathaṃ punaḥ svabuddhiguṇaḥ kriyate ?*
> *ayaṃ rāśiḥ kena guṇitedam ⟨edition reads guṇitam idam⟩*
> *agrāntaram prakṣipya viśodhya vā asya rāśeḥ śuddhaṃ bhāgaṃ*
> *dāsyatīti agrāntare kṣiptam| sameṣu kṣiptaṃ viṣameṣu śodhyam iti*

---

[101][Shukla 1976; p.132, lines 15 to 19]

sampradāyāvicchedād vyākhyāyate|

⟨As for⟩ "having a clever ⟨quantity⟩ for multiplier", the meaning is: having a multiplier according to one's own intelligence.

⟨Question⟩

But how is the multiplier according to one's own intelligence?

⟨It should answer this question:⟩ Will this quantity (the remainder), multiplied by what ⟨is sought⟩ give an exact division, when one has added or subtracted this difference of remainders ⟨to the product⟩?

⟨As for⟩ "Added to the difference of remainders"; ⟨it is⟩ added when ⟨the number of placed terms is⟩ even, subtracted when uneven, as it has been explained by an uninterrupted tradition.

From this remark, we can deduce the following computation.

If the number of placed terms is even ($n = 2p+1$, and, because the placement starts with the quotient q_2, the number of placed terms is $n - 1 = 2p$) one should solve the following equation having the following pair of integer unknowns: (k, l), where k is called "the clever ⟨quantity⟩" (*mati*).

$$l = \frac{r_n k + c}{r_{n-1}} = \frac{r_{2p+1} k + c}{r_{2p}},$$

where $c = R_1 - R_2$.

If the number of placed terms is not even ($n = 2p$, so that the number of placed terms is $n - 1 = 2p - 1$), the following equation should be solved:

$$l = \frac{r_n k - c}{r_{n-1}} = \frac{r_{2p} k - c}{r_{2p-1}}.$$

We do not know how these equations where solved. They have the same form as the problem solved by a pulverizer without remainder. However, only one solution is sought. It is not required that this solution is the smallest possible. The clever quantity, may have been found by trial and error.

Step 4

Sanskrit *Ab.2.33a. adhoparigunitam antyayug*

English Ab.2.33a. The one above is multiplied by the one below, and increased by the last.

Bhāskara furthermore adds:[102]:

---

[102][Shukla 1976; p.132 lines 20 to 23]

evaṃ paraspareṇa labdhāni padāny āsthāpya, matiś cādhaḥ,
paścimalabdhaś ca matyā adhaḥ| (...) evaṃ bhūyo bhūyaḥ karma
yāvat karma parisamāptitam iti|
When one has placed in this way the terms obtained by the mutual ⟨division⟩, the clever ⟨quantity⟩ is placed below, and the last
obtained below the clever ⟨quantity⟩. (...) In this way, again and
again the operation ⟨is repeated⟩ until the computation comes to
an end.

General comments The placement will then be:

$$q_2$$
$$q_3$$
$$\vdots$$
$$q_n$$
$$k$$
$$l$$

Then the operation:" The one above is multiplied by the one below, and
increased by the last ", is repeated, for all rows, beginning from the bottom
$(i = n, n - 1, ..., 2)$:

$$\begin{matrix} q_i \\ q'_{i+1} \\ q'_{i+2} \end{matrix} \quad \longrightarrow \quad \begin{matrix} q_i q'_{i+1} + q'_{i+2} \\ q'_{i+1} \end{matrix}$$

The third element from the bottom of the column is replaced by the result
of the computation prescribed, and the last element is deleted.

This procedure is repeated until only two elements remain.

$$q'_2$$
$$q'_3$$

(q'_2, q'_3) is a pair of integer solutions of the original problem[103], which is not
mentioned in the text. The procedure continues, considering q'_2, from which
another couple of solutions will be derived.

---

[103]Please see the last part of this section of the supplement BAB.2.32-33.

Step 5

Sanskrit *Ab.2.33b ūnāgracchedabhājite*

English Ab.2.33b. When ⟨the result of this procedure⟩ is divided by the divisor of the smaller remainder.

Bhāskara furthermore adds [104]:

> *ūnāgracchedhabhājite śeṣam, (. . .) pūrvagaṇitakarmaṇā*
> *niṣpannarāśer vibhaktaśeṣaṃ parigṛhyate|*
> ⟨As for⟩ "When ⟨the result of this procedure⟩ is divided by the divisor of the smaller remainder, the remainder". (. . .) of the division of, the quantity produced by means of the previous mathematical operation, by the divisor of the smaller remainder is understood.

General comments In other words, the solution, q'_2, is divided by b:

$$\frac{q'_2}{b} = t + \frac{s}{b} \Leftrightarrow q'_2 = bt + s \quad (o \le s < b).$$

The remainder, s, is thereafter considered. s is the least positive solution for x of the original problem[105], this is not mentioned in the text.

Step 6

Sanskrit *Ab.2.33bcd. śeṣam adhikāgracchedaguṇam dvicchedāgram adhikāgrayutam*

English Ab.2.33bcd. The remainder multiplied by the divisor of the greater remainder and increased by the greater remainder, is the ⟨quantity that has such⟩ remainders for two divisors.

Bhāskara furthermore adds[106]:

> *tad dvayor api chedayor bhājyarāśir bhavatīti|*
> . . . That is the quantity to be divided for (i.e. by) both of these two divisors.

General comments
$$N_1 = as + R_1.$$

N_1 is the least positive integer that satisfies the original problem, and, at the same time, it is regarded as the "remainder" (*agra*) corresponding to the two divisors, a and b, when there is another problem: Find the number N that when divided by ab leaves for remainder N_1, and when divided by another number leaves another given remainder.

---

[104][Shukla 1976; p. 132 lines 23 to 25]
[105]Please see the last part of this section of the supplement for BAB.2.32-33.
[106][Shukla 1976; p.133, lines 2-3]

X.2.2 Understanding the general case of the pulverizer with remainder

Let us recall that the problem treated ("What is the natural number N that divided by a leaves R_1 for remainder and divided by b leaves R_2 for remainder?"), can be summarized as follows:

$$N = ax + R_1 \quad 0 \leq R_1 < a$$
$$N = by + R_2 \quad 0 \leq R_2 < b$$

a   Preliminary remarks

a.1   Conditions on a and b   The original problem supposes that $a, b > 1$, since a division by 1 would leave no remainder, and that the problem if one of them were equal to zero would equally have no sense in this context.[107]

If $R_2 = R_1 = R$ when a and b are not coprime (that is their only common divisor is 1), as we can see in Example 4, then the smallest integer solution N would be

$$N = LCM(a, b) + R,$$

where $LCM(a, b)$ is the Least Common Multiple of a and b. This is the case of the five first quantities in example 4. We do not know, however, how Bhāskara proceeded in this case.

a.2   Conditions on the remainders   Usually, in examples, $R_1 \neq R_2$ and $R_1 \neq 0, R_2 \neq 0$.

Let us remark here that the above system of equations has a solution if and only if $R_1 - R_2$ is a multiple of the Greatest Common Divisor of a and b. Indeed, let (x_0, y_0) be a solution. Then:

$$R_1 - R_2 = by_0 - ax_0.$$

It is a common result of elementary number theory[108] that such a number is necessarily a multiple of the Greatest Common divisor of a and b. So that there should always be a common multiple for a, b, and $R_1 - R_2$.

---

[107] If we consider however the set of equations written above, let us suppose that: either a or b are equal to zero. If say a would be equal to zero, then we would have a value for N, R_1, that would verify the original problem, if and only if

$$R_1 = by + R_2$$

has an integer solution, that is if and only if $R_1 - R_2$ is a multiple of b.

[108] See for instance, [Gareth&Jones 1998; Proof of Theorem 1.8., p. 10]

If a and b are coprime, then for any difference of remainders solutions can be found. Bhāskara in the case of this interpretation of the pulverizer problem does not make any such remark on a and b. However concerning a pulverizer without remainder, such a fact is stated rather clearly, as we have noted in the section concerning this procedure below.

When $R_1 = R_2 = 0$, then N is a common multiple of both a and b. If (x_0, y_0) is the smallest solution of this set of equations then by definition, N is the Least Common Multiple of a and b.

Bhāskara at the beginning of example 14 writes:

> *kaścid rāśiḥ sūryasya nirapavartitabhūdivasair bhāgaṃ hriyamāṇaḥ*
> *śūnyāgraḥ, candrasyāpi śūnyāgraḥ eva saḥ|*
> Some quantity when divided by the reduced number of terrestrial days
> ⟨in a *yuga*⟩ for the sun, has a zero-remainder (*śūnyāgra*), just that ⟨same
> quantity when divided by the reduced number of civil days in a *yuga*⟩
> for the moon too has a zero-remainder.

He later exhibits as such a quantity, the Least Common Multiple of both numbers.

b   Understanding the procedure   In the following we will consider that $a, b > 1$ and that $R_1 > R_2$, $c = R_1 - R_2$.

The process is interrupted, it seems, when the remainder obtained is sufficiently small[109]. We can formalize the process in the following way (in exactly the same terms as in Step 1 and 2 of the procedure described in the commentary):

For an arbitrary n:

$$\frac{a}{b} = q_1 + \frac{r_1}{b} \quad\Leftrightarrow\quad a = bq_1 + r_1$$

$$\frac{b}{r_1} = q_2 + \frac{r_2}{r_1} \quad\Leftrightarrow\quad b = r_1 q_2 + r_2$$

$$\frac{r_1}{r_2} = q_3 + \frac{r_3}{r_2} \qquad r_1 = r_2 q_3 + r_3$$

$$\frac{r_2}{r_3} = q_4 + \frac{r_4}{r_3} \qquad r_2 = r_3 q_4 + r_4$$

$$\vdots$$

$$\frac{r_{n-2}}{r_{n-1}} = q_n + \frac{r_n}{r_{n-1}} \qquad r_{n-2} = r_{n-1} q_n + r_n.$$

---

[109]Bhāskara's contemporary, Brahmagupta, and all following known authors continue the process until zero is obtained as remainder, and therefore do not compute the "clever quantity".

By using this set of equations, the equation (*) can be rewritten as a set of two equations, (A, i) and (B, i), for $i = 1, \ldots, n$.

$$y = \frac{ax + c}{b} = \frac{(bq_1 + r_1)x + c}{b} = q_1 x + y_1 \quad \text{where}$$

$$y_1 = \frac{r_1 x + c}{b} \quad (A, 1),$$

$$x = \frac{by_1 - c}{r_1} = \frac{(r_1 q_2 + r_2)y_1 - c}{r_1} = q_2 y_1 + x_1 \quad \text{where}$$

$$x_1 = \frac{r_2 y_1 - c}{r_1} \quad (B, 1),$$

$$y_1 = \frac{r_1 x_1 + c}{r_2} = \frac{(r_2 q_3 + r_3)x_1 + c}{r_2} = q_3 x_1 + y_2 \quad \text{where}$$

$$y_2 = \frac{r_3 x_1 + c}{r_2} \quad (A, 2),$$

$$x_1 = \frac{r_2 y_2 - c}{r_3} = \frac{(r_3 q_4 + r_4)y_2 - c}{r_3} = q_4 y_2 + x_2 \quad \text{where}$$

$$x_2 = \frac{r_4 y_2 - c}{r_3} \quad (B, 2),$$

$$\vdots$$

$$\begin{cases} y_{p-1} = q_{2p-1} x_{p-1} + y_p \\ y_p = \frac{r_{2p-1} x_{p-1} + c}{r_{2p-2}} \end{cases} \quad (A, p)$$

$$\begin{cases} x_{p-1} = q_{2p} y_p + x_p \\ x_p = \frac{r_{2p} y_p - c}{r_{2p-1}} \end{cases} \quad (B, p)$$

$$\begin{cases} y_p = q_{2p+1} x_p + y_{p+1} \\ y_{p+1} = \frac{r_{2p+1} x_p + c}{r_{2p}} \end{cases} \quad (A, p+1)$$

etc.

Now, with the equation (B, p) is associated an even number of quotients (q_{2p}), and in the computation of x_p, c is subtracted.

With the equation $(A, p+1)$ is associated an uneven number of quotients (q_{2p+1}), and in the computation of y_{p+1}, c is added.

We can recognize here the computation of the clever quantity and the quotient that is associated to it, as in Step 3 of the algorithm.

If the number of quotients is uneven, the equation $(A, p+1)$ should be solved by trial and error; the solution, k, for x_p is called "the clever ⟨quantity⟩" (*mati*).

$$l = \frac{r_n k + c}{r_{n-1}} = \frac{r_{2p+1}k + c}{r_{2p}}.$$

If the number of quotients is even, the equation of (B,p) should be solved by trial and error; the solution, k , for y_p is called "the clever $\langle$quantity$\rangle$" (*mati*).

$$l = \frac{r_n k - c}{r_{n-1}} = \frac{r_{2p}k - c}{r_{2p-1}}.$$

Once a couple of solutions is found, by working the solutions backwards, one arrives at a solution x for (*).

Indeed, by solving the second equation of $(A, p+1)$ (resp. of (B, p)), one obtains a numerical value for both (x_p, y_{p+1}) (resp. of (x_p, y_p)), which in turn gives a value for y_p (resp. for x_{p-1}). With this value of y_p (resp. of x_{p-1}) the value of x_p (resp. for y_{p-1}) can be computed and so forth until we have obtained a value for (x_1, y_1), which gives a value for x.

In other words, by using the succession of equations, for example in the case of an uneven number of quotients:

$$\begin{aligned}
y_p &= q_{2p+1}x_p + y_{p+1} & (A, p) \\
x_{p-1} &= q_{2p}y_p + x_p & (B, p-1) \\
y_{p-1} &= q_{2p-1}x_{p-1} + y_p & (A, p-1) \\
&\quad\vdots & \\
x_1 &= q_4 y_2 + x_2 & (B, 1) \\
y_1 &= q_3 x_1 + y_2 & (A, 1) \\
x &= q_2 y_1 + x_1; &
\end{aligned}$$

one thus arrives at a solution for x.

Now in this succession of equations we can recognize the computations of Step 4, taking for example an even number of quotients:

$$
\begin{array}{ll}
q_2 & q_2 \\
q_3 & q_3 \\
\vdots \quad\longrightarrow & \vdots \quad\longrightarrow \cdots \\
q_{2p-1} & q_{2p-1} \\
q_{2p} & q_{2p}y_p + x_p = x_{p-1} \\
k = y_p & y_p \\
l = x_p &
\end{array}
$$

$$
\begin{array}{lll}
q_2 \quad\longrightarrow & q_2 & \longrightarrow \quad q_2' = q_2 y_1 + x_2 = x \\
q_3 & q_3' = q_3 x_1 + y_2 & \qquad\qquad q_3' = y_1 \\
q_4' = x_1 & x_1 & \\
q_5' = y_2 & &
\end{array}
$$

As we can see, only q_2 is needed to compute x, which may explain why there is no need to "set down" q_1.

Step 5, by dividing that very value of x by the "smaller divisor", and thereafter considering the remainder of the division, assures that the value found for x is the smallest possible. Step 6 replaces the value for x in the first equation:

$$N = ax + R_1,$$

So that N_1, the value obtained for N is such that

$$N_1 = as + R_1.$$

c   Procedure with more than two quantities and short cut N_1 satisfies the original problem, and, at the same time, it is regarded as the "remainder" (*agra*) corresponding to the two divisors, a and b, when there is another problem: Find a number N that when divided by ab leaves for remainder N_1. This can be formalized as

$$N = (ab)u + N_1.$$

A solution, N, of this problem is also such that when divided by a, it has for remainder R_1. Likewise, when N is divided by b, it has R_2 for remainder. This property is used when the problem concerns more than two couples of divisors and remainders. This is the case for instance in examples 3 and 4. If one has to solve a problem with more than two couples of divisors and remainders, if all the remainders are equal an evident solution will be the LCM of all divisors increased by the remainder (this is the case of the solution the example of Ms. E would bear). If just a certain number of these integers have the same remainder, the problem will be equivalent to solving the pulverizer of the LCM of those integers with their common remainder, and the others.

In Example 1, Bhāskara stops short of the "Euclidian Algorithm". The clever quantity he computes and the corresponding quotient, correspond, with our notations, to the computation of:

$$y_1 = \frac{r_1 x + c}{b} \ (A, 1).$$

The clever quantity is hence a value for x, which is then reduced to its smallest possible value by Step 5, and with which the value of N is computed in Step 6.

We will briefly expose here the steps followed by Bhāskara when he uses his short cut, and when considering more than two quantities.

X.2.3 Bhāskara's short cut

In Example 1, Bhāskara uses a "short-cut" whose steps we will now expose. The problem solved is the same and starts in the same way:

Step 1

"One should divide the divisor of the greater remainder by the divisor of the smaller remainder."

Supposing $R_1 > R_2$, then a is "the divisor of the greater remainder", and b is "the divisor of the smaller remainder":

$$\frac{a}{b} = q_1 + \frac{r_1}{b} \quad \Leftrightarrow \quad a = bq_1 + r_1.$$

However here r_1 is considered sufficiently "small" and step 2 is skipped

Step 3

The number of placed terms is considered to be even.

One should solve the following equation having the following pair of integer unknowns: (k, l), where k is called "the clever $\langle$quantity$\rangle$" (*mati*),

$$l = \frac{r_1 k + c}{b}.$$

Step 4 is skipped also but the "setting-down" would be: $\begin{matrix} k \\ l \end{matrix}$

Step 5

The upper element of this column, k is divided by b:

$$\frac{k}{b} = t + \frac{s}{b} \quad \Leftrightarrow \quad k = bt + s \quad (o \le s < b).$$

The remainder, s, is thereafter considered.

Step 6

$$N_1 = as + R_1.$$

N_1 is the least positive integer that satisfies the original problem.

X.2.4   Procedure for problems with more than two couples of numbers

Problem

What is the integer N that when divided by a_1 has r_1 for remainder, that when divided by a_2 has r_2 for remainder, $\cdots$, that when divided by a_n has r_n for remainder?

Procedure

A first pair of couples is chosen (say (a_1, r_1) and (a_2, r_2)) to which the pulverizer procedure is applied, and for which an integer N_1 is found. Then a following pair is taken (say, (a_3, r_3)) , to which the pulverizer procedure is applied together with the couple formed of the product of the previous divisors and the result found ($(a_1 a_2, N_1)$). And so forth, until all the couples are used. The last pulverizer procedure applied gives the solution of the problem. If two remainders are the same, Bhāskara indicates in Example 4:

> *atrecchayā 'dhikāgro rāśiḥ parikalpanīyaḥ|*
> In this case, the quantity which has the greater remainder should be chosen according to one's will.

We do not know if Bhāskara computed the largest common multiple of these divisors, in order to overcome the problem that occurs when two divisors are multiples of one another.

X.3   Procedure of the pulverizer without remainder

We will present here the different steps of this algorithm such as described in the general commentary. Then we will present two alternative procedures, solving the same problem, and found in the "procedure" part of solved astronomical examples.

X.3.1   General procedure

Problem

What is the integer x, that multiplied by a, increased or decreased by c and divided by b, produces an integer y?

In other words the problem consists of finding two integers (x, y) that verify

$$y = \frac{ax \pm c}{b}.$$

a, b and c are known positive integers. x is called the pulverizer or the multiplier (*guṇaka*), y the quotient (*labdha*).

In the "setting-down" part of examples, this is the pattern followed: $\begin{matrix} a & & c \\ b & & \end{matrix}$

Sometimes c is omitted.

At the beginning of Example 22 Bhāskara writes[110]:

> *bhāgahārabhājyāgrāmam ekena apavartanacchedena apavartaṇaṇ kṛtvā*
> *pūrvavat kuṭṭākāraḥ kriyate| atha punar etāni bhāgahārabhājyāgrāṇi*
> *chedenaikanāpavartanam na prayacchati yathā tathā sāv uddeśakaḥ,*
> *tādṛṣa's caiko rāśir eva nāsty ato na ānīyate|*
> When one has performed the reduction, by a unique reducing divisor, of
> the divisor, dividend and remainder, as before, a pulverizer is performed.
> Now, on the other hand, ⟨if⟩ that example is such that these divisor,
> dividend and remainder do not allow such a reduction with a unique
> divisor, as there is no such one quantity ⟨that satisfies this equation⟩,
> ⟨such a quantity⟩ is not computed ⟨with a pulverizer⟩.

So that as we have noted above, Bhāskara suggests reducing the numbers used in
examples before starting the computation (these truly get to huge proportions in
astronomical problems) but is also well aware that c should be a multiple of a and
b in order for such a problem to have a solution.

Step 1

Sanskrit *adhikāgrabhāgahāram chindyād ūnāgrabhāgahāreṇa*

English One should reduce the divisor which is a large number ⟨and the dividend⟩
by a divisor which is a small number.

General Comments In other words, one should discard common factors from a
(the dividend) and b (the divisor), a new couple (a', b') is therefore considered;
where a' and b' are coprime (that is their sole common divisor is 1). This
step can be seen as a "short-cut" for the following process of the "Euclidian
Algorithm". Practically, Bhāskara always discards their GCD.

Step 2–Step 4

As we have noted before, if we consider the problem solved by a pulverizer with
remainder: $R_1 > R_2$, and $R_1 - R_2 = c$,

$$\begin{cases} N = & ax+ & R_1 \\ N = & by+ & R_2 \end{cases} \Leftrightarrow y = \frac{ax+c}{b}$$

Therefore, as noted by Bhāskara as well, these steps are similar to Step 2- Step 4
of the pulverizer with remainder.

---

[110][Shukla 1976; last paragraph p.149-150]

Therefore here, the first division is that of the divisor *by* the dividend. In the end of this process we have two quantities, q'_2 and q'_3.

Step 5

Sanskrit *ūnāgracchedabhājite śeṣam*

English When ⟨the remaining upper quantity⟩ is divided by the divisor which is a small number, the remainder is ⟨the pulverizer. When the lower one remaining is divided by the dividend the quotient of the division is produced.⟩

Bhāskara further glosses[111]:

> *upari[rāśiḥ] bhāgahāreṇa bhaktaḥ [kāryaḥ], adhorāśir bhājya rāśinā bhājyaḥ*
>
> The upper [quantity should be made to be] divided by the divisor; the lower quantity should be divided by the dividend quantity. (...)

The two remainders are the pulverizer and the quotient of the division.

General With the same notation as before q'_2 ("the upper quantity") is divided by b ("the divisor"):

$$q'_2 = tb + u.$$

u is called the pulverizer.

q'_3 ("the lower quantity") is divided by a ("the dividend"):

$$q'_3 = va + w.$$

w is called the quotient.

The result is usually set down in a column: $\begin{array}{c} u \\ w \end{array}$

At the end of his resolution of Example 9[112], Bhāskara indicates:

> *[athavā] yāvad abhirūcitaṃ pṛcchakāya*
>
> [Or else] until it pleases the inquirer (*pṛcchaka*), ⟨the values should be increased by multiples of the constants⟩.

This somewhat elliptic remark, may refer to the following rule, given in the *Mahābhāskarīya* [Shukla 1960; sk p. 8, eng. p. 40]:

---

[111][Shukla 1976; p.135 lines 17 to 21]
[112][Shukla 1976; p.139]

> *prakṣipya bhāgahāraṃ kuṭṭākāre punaḥ punaḥ prājñaiḥ|*
> *yojyaṃ ca bhāgalabdhaṃ bhājye prastārayuktyaiva||*

Mbh.1.50. (To obtain the other solutions of a pulverizer) the intelligent (astronomer) should again and again add the divisor to the multiplier and the dividend to the quotient as in the process of *prastāra* ("representation of combinations").

In other words if (m, n) is a solution of

$$y = \frac{ax \pm c}{b},$$

where (x, y) are the unknowns, then, for any integer t,

$$m_t = m + tb$$
$$n_t = n + ta$$

are also solutions of this problem.

X.3.2 Alternative procedures

a The *sthirakuṭṭāka* In his commentary on Example 7, and then systematically in all resolutions after this one, when solving

$$y = \frac{ax \pm c}{b},$$

Bhāskara, instead of the usual procedure, proposes as an alternative to solve with the same procedure the following problem:

$$y' = \frac{ax' \pm 1}{b}.$$

The values found as solution are then used in a Rule of Three, with the following proportions:

$$1 : x' = c : x''$$
$$1 : y' = c : y''$$

The smallest values possible for x and y are found, by considering the remainders of the divisions of x'' by b, and of y'' by a.

This is known in later literature as the *sthirakuṭṭāka* (fixed-pulverizer).

The versified table that ends the *gaṇitapāda* gives the smallest possible solutions for problems of the type

$$y = \frac{ax - 1}{b},$$

using many different types of astronomical constants[113].

Solutions of

$$u = \frac{av + 1}{b}$$

may be easily derived from the type above, as

$$x = b - v$$
$$y = a - u$$

If no general rule is given by Bhāskara in his commentary, such a process is described in the *Mahābhāskarīya* [Shukla 1960; p. 32-33]:

> *Mbh.45. rūpaṃ ekam apāsyāpi kuṭṭākāraḥ prasādhyate|*
> *guṇakāro 'tha labdhaṃ ca rāśī syātām uparyadhaḥ||*
> *Mbh.46.ab. iṣṭena śeṣam abhihatya bhajed dṛḍhābhyāṃ*
> *śeṣam dināni bhagaṇādi ca kīrtyate 'tra|*
> Mbh.I.45-46ab. Alternatively, the pulverizer is solved by subtracting one (i.e., by assuming the residue to be unity). The upper and lower quantities (in the reduced chain) are the (corresponding) multiplier and quotient (respectively). By the multiplier and quotient (thus obtained) multiply the given residue, and then divide the respective products by the abraded divisor and dividend. The remainders obtained are here (in astronomy) the *ahargaṇa* and the revolutions (performed respectively).

This can be understood as follows:

If (m, n) is a solution of

$$y = \frac{ax \pm 1}{b},$$

where (x, y) are the unknowns. If (m_0, n_0) are respectively the remainders of the division of cm by b, and of cn by a,

$$m_0 = cm - bq \quad (0 \leq m_0 < b),$$
$$n_0 = cn - aq \quad (0 \leq n_0 < a),$$

then, (m_0, n_0) is a solution of

$$y = \frac{ax \pm c}{b}.$$

---

[113]We have not translated this versified table. It is summarized, and all values given, in [Shukla 1976; Appendix ii, p.335-339]

b   Another alternative   In his resolution of Example 11[114], Bhāskara describes an alternative procedure:

> *atra bhāgahāreṇa bhājyaṃ vibhajya labdhaṃ pṛthagavinaṣṭaṃ*
> *sthāpayet| śeṣasya bhūdivasānāṃ ca kuṭṭākāraṃ kṛtvā*
> *labdhasyoparirāśiṃ kuṭṭākāram avinaṣṭasthāpitena pṛthak*
> *saṃguṇayya bhāgalabdhaṃ prakṣipet| bhāgalabdhaṃ bhavati|*
> In this case, having divided the dividend by the divisor, one should place the quotient separately ⟨and keep it⟩ unerased. When one has performed the pulverizer of the terrestrial days and the residue, when one has multiplied separately the higher quantity of the ⟨two⟩ obtained by the pulverizer of the ⟨quantity⟩ kept unerased, one should add the quotient of the division ⟨which stands below⟩. ⟨This⟩ produces the quotient of the division.

Which can be understood as follows. What is obtained at the end of the process which proceeds upwards is

$$q'_2 = \frac{x}{y_1}$$

where y_1 is defined as

$$y = xq_1 + y_1.$$

Bhāskara, here indicates that one should set aside q_1 defined as the quotient of the division of a by b:

$$a = bq_1 + r_1.$$

Therefore the computation described here corresponds to a computation of y:

$$xq_1 + y_1 = y.$$

X.4   Astronomical applications

The kind of astronomical problem solved by the procedure of the pulverizer without remainder is introduced in Bhāskara's commentary without an explanation relating that process to given astronomical problems. These relations, however, can be found in the *Mahābhāskarīya*.

The basic idea is that the number of revolutions of a given planet, during a certain time is not a round number, but has, in addition to an integral value, a fractional part, or residue (*śeṣa*). This is also true, if are considered not only the number of revolutions, but also the number of signs (*rāśi* or *bhagaṇa*), degrees (*bhāga*) or

---

[114][Shukla 1976; p.141, line 15-18]

minutes (*liptā*), crossed by the planet during a given time. This time is usually evaluated in terms of civil days (*ahargaṇa*).

We will consider from now on, the following notations[115]:

Let A_y be the number of civil days in a *yuga*, G_y the number of revolutions performed by planet g in a *yuga*.

All the planet's revolutions in a *yuga* are given in Ab.1.3; the number of civil days in a *yuga* are deduced from both Ab.1.3 and Ab.3.3 and 5. This computation is described in the Appendix 4, which shows how this value of A_y is obtained: $A_y = 1577917500$.

As A_y and G_y will respectively be the dividend and divisor of a pulverizer without remainder, they are systematically reduced by their greatest common divisor. This can be seen in Bhāskara's commentary, at the beginning of the section on *maṇḍalakuṭṭākāra* (p. 135-136):

> *etāv ūnāgracchedārtham paraspareṇa bhājyau| ṣeṣam ūnāgracchedaḥ*
> These two should be divided by one another in order ⟨to obtain⟩ the divisor which is a smaller number. What remains is the divisor which is a smaller number...

Since the "divisor which is a smaller number" is, in this case, the greatest common divisor of the two first numbers, it appears that it was found by what is commonly called "the Euclidian Algorithm".

In the following, for the sake of convenience, we will also call A_y and G_y the numbers obtained after reduction. (G_y is usually called in secondary literature, the "revolution number" of the planet.)

Let A be the number of days elapsed since a given epoch (*ahargaṇa*). Here it is always the number of civil days elapsed since the beginning of the *Kaliyuga*.

Let G be the number of revolutions performed by a planet g in A days. G can be decomposed as the integral number of revolutions (*maṇḍala*) performed, M, the integral number of signs (*rāśi*), R, degrees (*bhāga*), B, and minutes (*liptā*), L crossed.

All the procedures use the ratio

$$\frac{A}{A_y} = \frac{G}{G_y}.$$

The reasoning followed in all the problems is basically the same, involving different ratios, according to the units considered, and occasionally a difference of sign in the pulverizer to solve, whether the fractional part of the path of g is considered

---

[115] All the notations used in this supplement are summed up on a list, at the end of this supplement.

as a surplus of the integral number of revolutions, or the part missing to obtain an integral number of revolutions. For the sake of simplicity, we have set aside here both the operations involving the reduction of the numbers of days and revolutions in a *yuga* and those converting values given in examples in homogeneous units (that is the conversion of a latitude given in degrees and minutes into minutes, etc.).

X.4.1 Planet's pulverizer (maṇḍalakuṭṭāka)

This computation concerns the commentary on verses 32-33, p.136-138. The planet considered is the sun.

a Planet's pulverizer with the residue of revolutions

Problem Let $A = x$, be the number of days elapsed since a given epoch (*ahargaṇa*), usually the beginning of the *Kaliyuga*. Let $M = y$ be the integral number of revolutions (*maṇḍala*) of a planet g during x days. These are the unknowns to be found, knowing:

-λ, the mean longitude of planet g in minutes after x days. ($\lambda = (30 \times 60)R + (60 \times B) + L$.)

- G_y, the reduced number of revolutions of planet g in a *yuga*.

- A_y, the reduced number of civil days in a *yuga*.

In the "setting down" part of examples, the disposition follows this pattern:

Integral number of signs crossed	Integral number of degrees crossed	Integral number of minutes crossed
R	B	L

or

Integral number of signs crossed	R
Integral number of degrees crossed	B
Integral number of minutes crossed	L

Procedure with the mean longitude Let λ be the mean longitude of planet g in minutes. R_M the "residue of revolutions", is defined as follows:

$$R_M = \frac{\lambda \times A_y}{21600}.$$

In the *Mahābhāskarīya*, the following rule occurs ([Shukla 1960; p. 33][116]):

---

[116]The first example given on this topic in Bhāskaraś commentary is explained in the pages 34-35.

> *rāśyādayo nirapavartitav āsaraghnā rāśyādimānabhajitāḥ*
> *pravadanti śeṣam*
> Mbh.1.46cd. (In the case the longitude of a planet is given in terms
> of signs, etc.) the signs, etc. are multiplied by the abraded number
> of civil days (in a *yuga*) and the product is divided by the number
> of signs, etc., (in a circle). The quotient is stated to be the residue
> (of revolutions).

In this case here the mean longitude of g (λ) is reduced to minutes, so that
the divisor is the number of minutes in a circle.

The residue of revolutions, R_M, can be understood as the number of civil
days taken to accomplish that part of a revolution indicated by λ_g. Since
21600 is the number of minutes in a circle, we have

$$\frac{R_M}{A_y} = \frac{\lambda}{21600}.$$

When computing R_M in his commentary, Bhāskara always considers an approximation of the quotient obtained, so that it may be an integer.

Two alternative methods are proposed having obtained this "residue of revolution", to solve the above problem:

Procedure 1 Find a couple solution of

$$y = \frac{G_y x - R_M}{A_y}.$$

$x = A$ is the number of days elapsed since a given epoch and $y = M$ is the
integral number of revolutions of a planet g during x days.

We can understand the process used here as the follows. When $\frac{\lambda_g}{21600}$, the
residual mean longitude in terms of revolutions, is the non-integer part of
the number of revolutions performed by G:

$$\frac{x}{A_y} = \frac{y + \frac{\lambda}{21600}}{G_y}.$$

This is equivalent to

$$y = \frac{G_y x - R_M}{A_y},$$

where $R_M = \frac{\lambda \times A_y}{21600}$.

Procedure 2 Uses a "*sthirakuṭṭāka*" process[117], that is:

Find a couple solution of
$$y' = \frac{G_y x' - 1}{A_y}.$$

The values obtained for this pulverizer are tabulated by Bhāskara at the end of the *gaṇitapāda*[118].

Then using the following ratios, x'' and y'' are computed:
$$1 : x' = R_M : x''$$
$$1 : y' = R_M : y'' \; ,$$

the smallest values possible for x and y are found, by considering the remainders of the divisions of x'' by A_y, and of y'' by G_y.

b Planet's pulverizer with the revolutions to be accomplished A similar procedure is found when considering the complementary part of the partial revolution accomplished. In this case, the part of the revolution to be crossed is added, when considering the pulverizer to solve.

Problem Let $A = x$ be the number of days elapsed since the beginning of the *Kaliyuga* (*ahargaṇa*). Let $M = y$ be the integral number of revolutions of a planet g during x days. These are the unknowns to be found, knowing:

-Δ, the part of a revolution to be accomplished by g so that the number of revolutions would be integer ($\lambda + \Delta = 1$ revolution).

- G_y, the reduced number of revolutions of planet g in a *yuga*.

- A_y, the reduced number of civil days in a *yuga*.

In the "setting down" part of examples, the disposition follows this pattern:

Integral number of signs to be crossed	R
Integral number of degrees to be crossed	B
Integral number of minutes to be crossed	L

A rule is given for this problem in the *Mahābhāskarīya*[119]:

> *gantavyam iṣṭaṃ yadi kasyacit syād gantavyayogād idam eva karma*|
> *rūpeṇa vā yojya vidhir vacintyaḥ sarvaṃ samānaṃ khalu lakṣaṇena*||
> Mbh.1.51. When the part (of the revolution) to be traversed by

---

[117]This process is explained in the section on the pulverizer without remainder.

[118]We have not translated this versified table. This table is summarized in [Shukla 1976; Appendix ii, p.335-339]

[119][Shukla 1960; sk p. 8-9, eng. p. 41]

some (planet) is the given quantity, then (also) the same process should be applied, treating the part to be traversed as the additive, or taking unity as the additive. All details of procedure are the same (as before).

Finding the part of a revolution to be accomplished The computation is exactly the same as the one described above. That is, if Δ is the part of a revolution to be accomplished by g in minutes, since 21600 is the number of minutes in a circle, then the " part of a revolution to be accomplished", R'_M, is:

$$R'_M = \frac{\Delta \times A_y}{21600}.$$

Having obtained this value two alternative methods are proposed to solve the above problem:

Procedure 1 Find the smallest couple solution of

$$y = \frac{G_y x + R'_M}{A_y}.$$

Procedure 2 Find the smallest couple solution of

$$y' = \frac{G_y x' + 1}{A_y}.$$

The values of

$$u' = \frac{G_y v' - 1}{A_y},$$

are tabulated by Bhāskara at the end of the *gaṇitapāda*. From these, x' and y' are obtained:

$$x' = A_y - v'$$
$$y' = G_y - u'$$

Then, using the same following ratios:

$$1 : x' = R'_M : x''$$
$$1 : y' = R'_M : y'' \ ,$$

the smallest values possible for x and y are found, by considering the remainders of the division of x'' by A_y, and of y'' by G_y.

X.4.2   Pulverizer with the residue of signs

Here, both the integral number of revolutions performed by g, M, and the following number of signs crossed by this planet, R, are unknown.

Problem Let $A = x$ be the number of days elapsed since the beginning of the *Kaliyuga* (*ahargaṇa*). Let $12 \times M + R = y$ be the integral number of signs crossed by g during x days. These are the unknowns to be found, knowing:

-λ', the remaining degrees and minutes crossed by g after x days in minutes ($\lambda' = 60 \times B + L$).

- G'_y, the reduced number of signs crossed by planet g in a *yuga*.

$$G'y = Gy \times 12,$$

as there are 12 signs in a revolution.

- A_y, the reduced number of civil days in a *yuga*.

In the "setting down" part of examples, the disposition follows this pattern, where the "0" indicates what is unknown or an empty space:

Integral number of revolutions crossed	0
Integral number of signs crossed	0
Integral number of degrees crossed	B
Integral number of minutes crossed	L

Finding the "residue of signs" A similar ratio to the one used in the cases above gives us the residue of signs (R_R), from λ', 1800 being the number of minutes in a sign:

$$\frac{R_R}{A_y} = \frac{\lambda'}{1800}.$$

In other words

$$R_R = \frac{\lambda' \times A_y}{1800}.$$

Having obtained the residue of signs three alternative methods are proposed to solve the above problem:

Procedure 1 Find the smallest couple solution of

$$y = \frac{G'_y x - R_R}{A_y}.$$

The value found for y is the number of signs crossed by g during x days. The remainder of the division of y by 12 will give the number of revolutions performed by g in x days.

Procedure 2 Find a couple solution of

$$y' = \frac{G'_y x' - 1}{A_y}.$$

These values are tabulated by Bhāskara at the end of the *gaṇitapāda*. Performing a Rule of Three with 1 and R_R, and dividing the results respectively by A_y and G'_y will give the results.

Procedure 3 Find a couple solution of

$$v' = \frac{12u' - 1}{A_y}.$$

The following procedure is not given by Bhāskara, thought he indicates that a Rule of Three should be used. We can consider the following, thought this is just a hypothetical construction in order to understand why this pulverizer is computed:

We have the ratio

$$\frac{\lambda}{21600} = \frac{R_M}{A_y},$$

where, as in section C.3.1, R_M is the residue of revolutions and $\lambda = (30 \times 60)R + (60 \times B) + L = (30 \times 60)R + \lambda'$. So this is equivalent to

$$\frac{(30 \times 60)R + \lambda'}{21600} = \frac{R_M}{A_y}.$$

Now if we consider this residual part of revolutions accomplished, not in terms of minutes, but in terms of signs (or if we reduce the left-hand fraction by $30 \times 60 = 1800$) we have

$$\frac{R + \frac{\lambda'}{30 \times 60}}{12} = \frac{R_M}{A_y}.$$

Let $v = R$ and $u = R_M$ and we recognize here:

$$v = \frac{12u - \frac{\lambda' \times A_y}{1800}}{A_y} = \frac{12u - R_R}{A_y}.$$

Bhāskara would thus solve this problem by a *sthirakuṭṭāka*.

u being the residue of revolutions, the problem

$$y' = \frac{G_y x - u}{A_y},$$

when solved gives with x the number of days elapsed since a given epoch, and with y' the number of revolutions accomplished in x days. Together with the value found for v, we can find the total number of signs crossed by g in x days.

a   Pulverizer for the residue of degrees   The process follows the same pattern as before, the difference being that one seeks the total number of degrees crossed by g in x days, that is that, M, R and B are unknown.

Problem   Let $A = x$ be the number of days elapsed since a given epoch (*ahargaṇa*). Let $12 \times 30M + 30 \times R + B = y$ be the integral number of degrees crossed by g during x days. These are the unknowns to be found, knowing:

-$\lambda''_g = L$, the remaining minutes crossed by g after x days.

- G''_y, the reduced number of degrees crossed by planet g in a *yuga*.

$$G''y = Gy \times 360,$$

as there are 360 degrees in a revolution.

- A_y, the reduced number of civil days in a *yuga*.

In the "setting down" part of examples, the disposition follows this pattern, where the "0" indicates what is unknown or an empty space:

Integral number of revolutions crossed	0
Integral number of signs crossed	0
Integral number of degrees crossed	0
Integral number of minutes crossed	L

Finding the"residue of degrees" A similar ratio to the one used in the cases above gives us the residue of degrees (R_B), from λ_g'', 60 being the number of minutes in a degree:

$$\frac{R_B}{A_y} = \frac{\lambda_g''}{60}.$$

In other words

$$R_B = \frac{\lambda_g'' \times A_y}{60}.$$

Having obtained the residue of degrees three alternative methods are proposed to solve the above problem:

Procedure 1 Find the smallest couple solution of:

$$y = \frac{G_y'' x - R_B}{A_y}.$$

The value found for y is the number of degrees crossed by g during x days. The remainder of the division of y by 360 will give the number of revolutions performed by g in x days.

Procedure 2 Find a couple solution of

$$y = \frac{G_y'' x - 1}{A_y}.$$

These values are tabulated by Bhāskara at the end of the *gaṇitapāda*.

Performing a Rule of Three with 1 and R_B, and dividing the results respectively by A_y and G_y'' will give the required results. The remainder of the division of y by 360 (i.e. the number of degrees in a revolution) will give the number of revolutions performed by g in x days.

Procedure 3 Find a couple solution of

$$v' = \frac{30u' - 1}{A_y}.$$

The following procedure is not given by Bhāskara, though he indicates that a Rule of Three should be used. We can consider the following:

We have the ratio

$$\frac{\lambda'_g}{1800} = \frac{R_R}{A_y}$$

which is equivalent to:

$$\frac{(60 \times B) + L}{1800} = \frac{R_R}{A_y}.$$

Now if we consider this residual part of signs crossed, not in terms of signs but in terms of degrees (or if we simplify the left-hand fraction by 60):

$$\frac{B + \frac{L}{60}}{30} = \frac{R_R}{A_y}.$$

Let $v = B$ and $u = R_R$, then

$$v = \frac{30u - \frac{\lambda''_g \times A_y}{60}}{A_y} = \frac{30u - R_B}{A_y}.$$

Since u is residue of signs, the problem

$$y' = \frac{G'_y x - u}{A_y},$$

when solved, gives with x the number of days elapsed since the beginning of the *Kaliyuga*, and with y' the number of revolutions accomplished and the number of signes crossed in x days. Together with the value found for v, we can find the total number of degrees crossed by g in x days.

b   Pulverizer for the residue of minutes   The procedure follows the same pattern, considering residual seconds, crossed by G.

X.4.3   Week-day pulverizer

Problem A planet g, has a given mean longitude, λ, on a week day V. After a certain number of weeks (w) and a couple of days (a), g has the same longitude on another week-day, V_a.

Let a be the number of week-days seperating V from V_a (V excluded, V_a included; $a \leq 7$).

Let A_V be the number of days elapsed in the *Kaliyuga* when the sun is in V.

Let A_{V_a} be the number of civil days elapsed in the *Kaliyuga* for which the sun on V_a has the given mean longitude in V.

A_V and A_{V_a} are to be found, knowing λ on V; A_y and G_y.

Resolution The computation of A_V corresponds to a usual "planet-pulverizer": If $A_V = x$ and y=M, then by solving with a pulverizer the problem:

$$y = \frac{G_y x - R_M}{A_y},$$

the required value for A_V is found. Let x_0 be such a value.

In the *Mahābhāskarīya* there is the following rule[120]:

> *apavartitav āsarādiśeṣāt kramaśastān apanīya rūpapūrvam|*
> *kuṭṭākalabdharāśim eṣāṃ guṇakāraṃ samuśanti vārahetoḥ||*
> MBh.1.48. Divide the abraded number of civil days (in a *yuga*)
> by 7. Take the remainder as the dividend and 7 as the divisor.
> Also take the excess 1,2, etc., of the required day over the given
> day as the residue. Whatever number (i.e. multiplier) results on
> solving this pulverizer is the multiplier of the abraded number of
> civil days. The product of these added to the *ahargaṇa* calculated
> (for the given day) gives the *ahargaṇa* for the required day.

And in his introduction to Example 12 of the commentary to verses 32-33, Bhāskara writes:

> *nirapavartitabhūdineṣu saptahṛtāvaśiṣṭeṣu kuṭṭākāraḥ kriyate|*
> *grahavāro yo nirdiṣṭas tasmād y[ad u]ttaro grahavāras tataḥ*
> *prabhṛti ekottarayā vṛddhyāpacayaṃ parikalpya evaṃ labdhaṃ*
> *kuṭṭākāro nirapavartitabhūdinānāṃ guṇakāras tena guṇiteṣu*
> *nirapavartitabhūdineṣu nirdiṣṭasūryeṇānītam ahargaṇaṃ*
> *prakṣipya jātadivasatulyaḥ kāla ādeṣṭavyaḥ*
> A pulverizer should be performed for the residue of the division
> by seven of the reduced terrestrial days. When one has chosen a
> subtractive ⟨term for the pulverizer⟩ by means of a one-by-one in-
> crease beginning with the weekday which is immediately after the
> indicated week-day, what is obtained in this way is the pulverizer
> which is the multiplier of the reduced terrestrial days; when one
> has added the passed number of days ⟨in the *Kaliyuga*, obtained
> with⟩ the indicated sun, to the reduced terrestrial days multiplied
> by that ⟨pulverizer⟩, the time equal to what has been produced
> should be announced ⟨as the answer.

In this case, the pulverizer considered is, if A'_y is the residue of the division of A_y by seven ($A'_y = A_y - 7q$), a corresponding to the "one-by-one increase beginning with the weekday which is immediately after the indicated week-days":

---

[120][Shukla 1960; sk p. 8, eng. p.36-37(this is an adaptation – see note 1, p.37)]

$$w' = \frac{A'_y v' - a}{7}.$$

If (v'_0, w'_0) is a solution, then

$$A_{V_a} = A_y v'_0 + x_0.$$

This can be understood as follows: if A_y is the reduced number of civil days in a *yuga*, so that the number of weeks in a *yuga* is $\frac{A_y}{7}$, then we have the proportion:

$$\frac{A_{\Delta V}}{A_y} = \frac{w + \frac{a}{7}}{\frac{A_y}{7}},$$

where $A_{\Delta V}$ is the number of civil days after which the sun, having had that given longitude in V, has the same longitude in V_a, and w is the number of weeks in $A_{\Delta V}$, so that $A_{\Delta V} = 7w + a$.

If[121] $v = \frac{A_{\Delta V}}{A_y}$, then we have

$$\frac{v}{7} = \frac{w + \frac{a}{7}}{A_y}.$$

From this proportion we can deduce the following problem solved by a pulverizer:

$$w = \frac{A_y v - a}{7}.$$

Let (v_0, w_0) be a solution of that problem.

Since A_V is the number of days elapsed in the *Kaliyuga* when the sun is in V, A_{V_a} the number of civil days elapsed in the *Kaliyuga* for which the sun on V_a has the given mean longitude, and $A_{\Delta V}$ the number of civil days after which the sun, having had that given longitude in V, has the same longitude in V_a then

$$A_{V_a} = A_V + A_{\Delta V}.$$

---

[121]There seems to be a paradox here, as $A_{\Delta V}$ is thus defined as a multiple of A_y, therefore $A_{\Delta V} > A_y$. This assumption without any comment is also made by K.S. Shukla, when he solves example 12. [Shukla 1976; p.317] (A being what we denote $A_{\Delta V}$, 210389 being the reduced number of civil days in a *yuga* for the sun). We can, nonetheless, remark that A_y is, here, the *reduced* number of terrestrial days in a *yuga* and not the total number, so that this is not as absurd as it may seem. However, just why should this be presupposed and whether this is the exact rending of the computation described by Bhāskara, remains to be investigated.

By definition of v_0, and x_0:

$$A_{V_a} = x_0 + A_y v_0.$$

Now the particular solution, v_0', for v' makes also the quotient

$$\frac{A_y v - a}{7}$$

integer because

$$w' + 7q = \frac{A_y v' - a}{7}.$$

X.4.4 Particular pulverizers

Some of the examples proposed by Bhāskara combine several of the problems and procedures exposed above.

a A particular planet's pulverizer The problem here considers the remaining part of a degree to be crossed by a planet, combining thus a "pulverizer for a revolution to be accomplished" and "a pulverizer with the residue of degrees". In Example 13 [Shukla 1976; p.143] is exposed a problem and resolution of this type.

Problem Let $A = x$ be the number of days elapsed since a given epoch (*ahargaṇa*). Let $(12 \times 30)M + 30R + B = y$ be the integral number of degrees crossed by g during x days. These are the unknowns to be found, knowing:

-Δ'', the part of a degree to be crossed by g so that the number of degrees crossed since the beginning of the *Kaliyuga* would be integer.

- G_y'', the reduced number of degrees crossed by planet g in a *yuga*.

$$G''y = 360 \times Gy,$$

as there are 360 degrees in a revolution.

- A_y, the reduced number of civil days in a *yuga*.

Procedure After having computed the residue of degrees to be crossed,

$$R_B' = \frac{\Delta'' \times A_y}{60},$$

the following problem is to be solved directly by a pulverizer procedure, or by using a *sthirakuṭṭāka*:

$$y = \frac{G''_y x + R'_B}{A_y}.$$

The value found for $y - 1$, when divided by 360 gives the integral number of revolutions performed by g in x days.

b   A particular week-day pulverizer

Problem In this case, the mean longitude of planet g_1 (λ_1), and the mean longitude of planet g_2 (λ_2) are known, for a given week-day (V); the number of days until they will both be of the same longitude again on another week-day (V_a) is what is sought.

Finding the LCM Let A_1 be the reduced number of days in a *yuga* for g_1; A_2 the reduced number of days in a *yuga* for g_2. The Lowest Common Multiple of these two numbers ($LCM(A_1, A_2)$), can be defined as:

$$LCM(A_1, A_2) = \frac{A_1 \times A_2}{GCD(A_1, A_2)}.$$

It is found by the following process:

-The Greatest Common divisor ($GCD(A_1, A_2)$) is found, probably by a "Euclidian algorithm".

In the case of the preliminary part of Example 14, it is defined as the quantity which leaves a zero remainder (*śūnyāgra*), when divided by A_1 or by A_2. It bears the name "⟨quantity⟩ having such remainder for two divisors." (*dvicchedāgra*).

-The quotient of the division of A_1 (resp. A_2) by $GCD(A_1, A_2)$ (q_1) (resp. q_2) is considered.

Then

$$LCM(A_1, A_2) = A_1 \times q_2 = A_2 \times q_1.$$

This is expressed quite elliptically in the preliminary part of Example 14, but corresponds to the computations carried out:

> *dvicchedāgrasaṃvargo hi nāma sadṛśikaraṇam*
> the product of ⟨one reduced day by the quotient of the other by the quantity⟩ having such remainder for two divisors (*dvicchedāgrasaṃvargo*) has the name "procedure of equalizing (*sadṛśīkaraṇaṃ*) for two quantities".

Finding the number of days elapsed in the Kaliyuga when g_1 and g_2 are in V

This involves a usual planet-pulverizer: The smallest integral solution found for x (x_0) in any of these equations gives the desired value

$$\begin{cases} y = \frac{G_1 x - R_{M_1}}{A_1}, \\ y = \frac{G_2 x - R_{M_2}}{A_2}. \end{cases}$$

A week-day pulverizer The following problem is solved by a pulverizer:

$$w = \frac{LCM(A_1, A_2)v - a}{7}.$$

Let v_0 be the smallest integral value found. Then

$$A_{\Delta_V} = LCM(A_1, A_2)v_0 + x_0.$$

Thus , the following equality explains this formulation of the problem:

$$\frac{w + \frac{a}{7}}{LCM(A_1, A_2)} = \frac{v}{7},$$

where

$$v = \frac{A_{\Delta V}}{LCM(A_1, A_2)}.$$

X.4.5   A pulverizer using the sum of the longitudes of planets

Problem Let $A = x$ be the number of days elapsed in the *Kaliyuga*. This is the unknown to be found, knowing:

-$\Sigma\lambda$, the sums of the mean longitudes of n planets, in minutes, after x days. $(\Sigma\lambda = \sum_{i=1}^{n}\lambda_i = \sum_{i=1}^{n}(30 \times 60)R_i + (60 \times B_i) + L_i, n \leq 7)$[122].

- ΣG_y, the reduced sum of the number of revolutions performed by each planet in a *yuga*.

- A_y, the reduced number of civil days in a *yuga*.

---

[122] A list of the planets is given in Ab.3.15.

Procedure The procedure, with these constants, is the same as in a regular planet's pulverizer. Having computed the residue of revolution of the sun,

$$\Sigma R_M = \frac{\Sigma\lambda \times A_y}{21600},$$

the problem to be solved by a pulverizer or by a *sthirakuṭṭāka* is

$$y = \frac{\Sigma G_y x - \Sigma R_M}{A_y}.$$

The smallest solution found for y is the sum of the revolutions performed by n planets in x days.

As before, the constant ratio behind this problem is

$$\frac{A}{A_y} = \frac{G_1}{G_{g_1}} = \cdots = \frac{G_n}{G_{g_n}},$$

so that

$$\frac{A}{A_y} = \frac{\Sigma G_y}{\Sigma G}.$$

This procedure is described in example 15 [Shukla 1976; p.144sqq]; where only two planets are considered, the sun and the moon. However Bhāskara adds:

> *evam anyeṣām api samāsapraśneṣu kuṭṭākāraḥ kalpanīyaḥ,*
> *rāśibhāgaliptāśeṣvapi| evam eva tricatuḥsamaseṣvapi vistareṇa vyākhyeyam|*

In this way, in questions concerning the sums of other ⟨planets⟩ too, a pulverizer is to be performed (*kalpanīya*), and also ⟨in questions⟩ concerning residues of signs, degrees and minutes. In this very way, in the case of the sums of three or four ⟨planets⟩ also an explanation should be given in detail ⟨if necessary⟩.

X.4.6 Knowing the number of revolutions performed by two planets

Problem The number of revolutions performed since the beginning of the *Kaliyuga* by g_1 (y) and the integral number of revolutions performed by g_2 (z) are sought, knowing:

-λ_2, the mean longitude of g_2 in minutes, known when g_1 completes a revolution.

-G_1 and G_2, (previously reduced by their greatest common divisor), the reduced sum of revolutions performed by g_1 and g_2 in a *yuga*.

Resolution The problem to be solved by a pulverizer without remainder or by a
sthirakuṭṭāka is

$$z = \frac{G_2 y - R_{M_2}}{G_1}.$$

This is understood by the following reasoning: If A is the number of civil
days elapsed at a given time, A_y the number of civil days in a *yuga*, then we
have:

$$\begin{cases} \dfrac{A}{A_y} = \dfrac{M_1 + \frac{\lambda_1}{21600}}{G_1} \\[2em] \dfrac{A}{A_y} = \dfrac{M_2 + \frac{\lambda_2}{21600}}{G_2} \end{cases}$$

And therefore

$$\frac{M_1 + \frac{\lambda_1}{21600}}{G_1} = \frac{M_2 + \frac{\lambda_2}{21600}}{G_2},$$

with the notation adopted above, that is

$$\frac{y}{G_1} = \frac{z + \frac{\lambda_2}{21600}}{G_2}.$$

From this equality the problem to be solved by a pulverizer is readily deduced.
Similarly, if the ratio considered for g_1 is measured in minutes then

$$\frac{21600 M_1 + \lambda_1}{G_1''} = \frac{M_2 + \frac{\lambda_2}{21600}}{G_2},$$

and the problem to be solved by a pulverizer would then be

$$z = \frac{G_2 Y - R_{M_2}}{G_1''},$$

where $Y = 21600y$, is the number of minutes crossed by g_1 since the beginning
of the *Kaliyuga*.

The problem and method to solve such a pulverizer is described in general terms by Bhāskara in this way[123]:

atha kaścid divasakaramaṇḍalaśeṣaparisamāptikāle janitaṃ
divicaramuddiśya divasakaraṃ divicarabhagaṇān pṛcchati, tasyāyam
upāyaḥ nirdiṣṭadivicaraṃ ravibhagaṇāṃścāpavartya kuṭṭākāro yojyaḥ|
Now, when pointing at ⟨the longitude of⟩ a planet produced at the time when the sun completes what remains of a revolution, someone asks the ⟨number of⟩ revolutions ⟨performed⟩ by ⟨that planet⟩, this is a method for that ⟨question⟩ -When one has reduced the ⟨number of⟩ revolutions ⟨performed⟩ by a planet ⟨in a *yuga*⟩ and the ⟨number of⟩ revolutions ⟨performed⟩ by the sun ⟨in a *yuga*⟩, a pulverizer should be applied.

He then proceeds to solve the problem given in example 16, and concludes by the following statement[124]:

athavā graham uddiśya graham evānyaṃ [pṛcchati tatr]āpi
bhāgahārabhājyaparikalpanayā kuṭṭākāraḥ kalpanīyaḥ|
Or else when ⟨someone⟩ pointing at a planet asks ⟨the number of passed revolutions⟩ of another planet only, then again a pulverizer should be performed by choosing ⟨an appropriate⟩ divisor and dividend.

Here therefore Bhāskara does not stress the unit in which the number of elapsed revolutions are obtained.

Mbh.1.10 gives the following procedure[125]

niśākaraṃ vā graham uccam eva vā kalīkṛtaṃ tat saha yātamaṇḍalaiḥ|
yatheṣṭanakṣatragaṇair hataṃ haret tadīyanakṣatraganais tataḥ kalāḥ||
10. The (mean) longitude of the moon, the planet, or the *ucca* (whichever is known) together with the revolutions performed should be reduced to minutes. The resulting minutes should then be multiplied by the revolution-number of the desired planet and (the product obtained should be) divided by the revolution-number of that (known) planet. The result is (the mean longitude of the desired planet) in terms of minutes.

In fact Example 16 of BAB.2.32-33 follows a computation in terms of revolutions whereas Example 17 follows the above rule given in the *Māhabhāskarīya*.

---

[123][Shukla 1976; p.145, line 16 sqq]
[124][Shukla 1976; p.146, line 13 sqq]
[125][Shukla 1960; p.2-3 skt, p. 7 eng.]

X.4.7 Time-pulverizer (velākuṭṭākāra)

In this case, the number of days elapsed since the beginning of the *Kaliyuga* is not integral: the longitude of planet g is not given at sunrise – a day is defined from one sunrise to another in this treatise – but at another time of the day: midnight, noon, or sunset[126].

Problem The integral number of days elapsed since the beginning of the *Kaliyuga* (x) and the number of revolutions performed by g in that time (y) are sought, knowing λ the mean longitude of g at a fractional part of the day (day$\pm\frac{1}{m}$, $2 \leq m \leq 4$), G_y and A_y.

Procedure The problem to be solved by a pulverizer without remainder or a *sthirakuṭṭāka* is

$$y = \frac{\frac{G_y}{m} \times X - R_M}{A_y},$$

where y is the number of revolutions performed by g in $x \pm \frac{1}{m}$ days and $X = mx \pm 1$.

If $\frac{1}{m}$ is subtractive ($\frac{X}{m} = x - \frac{1}{m} \Leftrightarrow X = mx - 1$), then the integral value of days elapsed since the beginning of the *Kaliyuga* is $x - 1$. Therefore the value sought is $x - 1 = \frac{X+1}{m} - 1$.

If $\frac{1}{m}$ is additive ($\frac{X}{m} = x + \frac{1}{m} \Leftrightarrow X = mx + 1$) then $x = \frac{X-1}{m}$ should be computed to obtain a solution.

The problem exposed in words here can be algebrised, in regard to a regular planet-pulverizer in this way:

$$y = \frac{G_y(x \pm \frac{1}{m}) - R_M}{A_y} \Leftrightarrow y = \frac{\frac{G_y}{m}(mx \pm 1) - R_M}{A_y}.$$

Bhāskara does not in fact describe exactly such a computation, concerning the passing first, from the pulverizer considering x to the one considering X and then from the result obtained for X to the one giving x.

In the part preceding Example 19, Bhāskara writes[127]:

> *kaścit graham udayakālād anyakālajanitaṃ pradaśyam divasagaṇam*
> *pṛcchati, tasyāyam ānayanopāyaḥ: iṣṭakālacchedaguṇitān*
> *nirapavartitabhūdivasān kṛtvā pūrvavat kuṭṭākāraṃ niṣpādya*
> *iṣṭakālachedhabhakto 'hargaṇaḥ|*

---

[126]Other subdivisions of the days can be also considered: this is indicated by Bhāskara in the part just before Example 21 which considers a fractional part of a day in *nāḍīs* (1/60th of a day).
[127][Shukla 1976; p. 147, line 15-17]

When someone pointing at ⟨the mean longitude of⟩ a planet produced
at a time different from sunrise, asks the number of days ⟨elapsed in the
Kaliyuga⟩, this is a method of computation for that ⟨question⟩:When
one has multiplied the reduced ⟨number of⟩ days ⟨in a *yuga*, for that
planet⟩ by the denominator of the desired time, and brought about a
pulverizer, as before, ⟨the pulverizer⟩ is divided by the denominator of
the desired time is the number of days ⟨elapsed in the *Kaliyuga*⟩.

Bhāskara, quite typically since he is summing up a general case, is elliptic concerning the computation of the integral number of days elapsed since the beginning
of the *Kaliyuga*. The first step he describes, that of multiplying by m a "reduced
number of days" has continued to be not understood. He states this again in
the "procedure" part of solved examples, but with no numerical illustration. This
may be referring to the computation $X = mx \pm 1$, however why then x would bear
such a name remains unclear. Secondly, repeatedly the passing from the pulverizer
obtained to the result sought (the integral number of days elapsed since the beginning of the *Kaliyuga*) is stated as a simple "division by the denominator of the
desired time", no other computation being stated. We note also that the integral
part of $\frac{X}{m}$ will give the value of $x - 1$ if m is subtractive, and the value of x if m
is additive. Therefore, this may have been the computation carried out here.

To sum it up, probably the computation we have algebrised in this case does not
render the exact steps followed by Bhāskara.

X.4.8 Finding the Residue of revolutions and a certain number of days, for two planets

This problem combines two pulverizers. Such a procedure may be seen in Example
23, where the two planets considered are the sun and Mars.

Problem Two planets g_1 and g_2 are considered. A certain amount of days, N is
sought, knowing that divided by A_1 (the reduced number of days in a *yuga*
for g_1) it leaves a remainder r_1 whose value is unknown, and divided by A_2,
it leaves a remainder r_2 whose value is unknown.

We can recognize here a problem that can be solved by a "pulverizer with
remainder" procedure, when r_1 and r_2 are known:

$$N = A_1 q_1 + r_1,$$

$$N = A_2 q_2 + r_2.$$

The values of r_1' and r_2' are known, and defined as

$$\frac{G_1 r_1}{A_1} = q_1' + \frac{r_1'}{A_1},$$

$$\frac{G_2 r_2}{A_2} = q_2' + \frac{r_2'}{A_2},$$

where G_1 and G_2 respectively are the reduced number of revolutions performed in a *yuga* by g_1 and g_2.

Procedure The last problem is equivalent to this one:

$$q_1' = \frac{G_1 r_1 - r_1'}{A_1},$$

$$q_2' = \frac{G_2 r_2 - r_2'}{A_2},$$

so that values of r_1 and r_2 may be found by means of one of the procedures for a "pulverizer without remainder".

r_1 (resp. r_2) is interpreted as the number of days elapsed since the beginning of the *Kaliyuga*; q_1' (resp. q_2') as the integral number of revolutions performed by g_1 (resp. g_2) during that time, and r_1' (resp. r_2') as the residue of revolutions, R_{M_1} (resp. R_{M_2}).

Having obtained r_1 and r_2, N is found by applying a second pulverizer.

X.4.9 Planetary pulverizer with several planets using orbital computations

This is the last type of problem illustrated by Bhāskara (in Examples 24-26), it combines a planetary pulverizer and the computations linking the length of the orbit of a planet to its mean longitude for a given number of elapsed days since the beginning of the *Kaliyuga*.

a Residues in respect to a planet's orbit   Let λ be the mean longitude of a given planet g.

$$\lambda = (M, R, B, L, S),$$

where M is the integer number of revolutions (*maṇḍala*) performed by the planet since the beginning of the *Kaliyuga*; R the remaining integer number of signs (*rāśi*) crossed, B the remaining integer number of degrees (*bhāga*) crossed, L the remaining integer number of minutes (*liptā*) crossed, and S, the remaining (*śeṣa*) fractional part of minutes crossed by that planet.

In terms of revolutions,

$$\lambda = M + \frac{R}{12} + \frac{B}{12 \times 30} + \frac{L}{12 \times 30 \times 60} + \frac{S}{12 \times 30 \times 60 \times (K \times A_y)}.$$

The residue of revolutions in respect to the planet's orbit is

$$Rk_M = \frac{R}{12} + \frac{B}{12 \times 30} + \frac{L}{12 \times 30 \times 60} + \frac{S}{12 \times 30 \times 60 \times (K \times A_y)}.$$

The residue of signs in respect to the planet's orbit is

$$Rk_M = \frac{B}{30} + \frac{L}{30 \times 60} + \frac{S}{30 \times 60 \times (K \times A_y)}.$$

The residue of degrees in respect to the planet's orbit is:

$$Rk_M = \frac{L}{60} + \frac{S}{60 \times (K \times A_y)}.$$

b   Case with two planets using a Residue of revolutions in respect to the planet's orbits

Problem The number of days elapsed since the beginning of the *Kaliyuga* and the mean longitudes, at that time, of two planets: λ_1, λ_2, are sought knowing:

-K_k the length in *yojanas* of the "orbit of the sky" (*khakakṣyā*) – the circumference of a great circle of the celestial sphere),

-K_1, K_2 the length in *yojanas* of the "orbit of the planets",

-A_y, the number of terrestrial days in a *yuga*,

-Rk_{M_1}, Rk_{M_2} the residue of revolutions of each planets at that time, in respect to the planet's orbit.

Orbital computations In the resolution of Example 24, Bhāskara quotes the following rule:

> *kakṣyābhir grahānayane khakaṣyāyā ahargaṇo guṇakākraḥ,*
> *svakaṣyābhūdinasaṃvargo bhāgahāra iti*
> In a computation of ⟨the mean longitude of⟩ planets by means of the orbits, the number of days ⟨elapsed in the *Kaliyuga*⟩ is a multiplier of the orbit of the sky, the divisor is the product of the terrestrial days ⟨in a *yuga*⟩ with its (the planet's) own orbit

In other words, for any planet:

$$\lambda_i = \frac{Kx}{A_y \times K_i}.$$

So that for our two planets we have

$$Kx = A_y \times K_1 \times \lambda_1 = A_y \times K_2 \times \lambda_2 = N.$$

Procedure Bearing the above equality in mind, for any planet:

$$A_y \lambda_i K_i = A_y K_i M_i + Rk_{M_i}.$$

In this problem M_i is sought and Rk_{M_i} is known.

The above equality may be written as a system of equations:

$$\begin{cases} N = A_y K_1 y + Rk_{M_1} \\ \\ N = A_y K_2 z + Rk_{M_2} \end{cases} \Leftrightarrow \quad z = \frac{A_y K_1 y - (Rk_{M_2} - Rk_{M_1})}{A_y K_2}$$

where y is the integral number of revolutions performed by the first planet and z the integral number of revolutions performed by the second planet.

This problem may be solved by a "pulverizer with remainder" procedure. Any one value found for y or z thus gives a value for N.

As Bhāskara states in the resolution of example 24:

> *pūrva likhitadvicchedāgrarāśir apavartitakhakakṣyāharganasaṃ-*
> *vargaity ataḥ svabhāgahārābhyāṃ vibhajya labdhaṃ*
> *sūryācandramasor yātabhāganāḥ*
> Since the previously written quantity that has ⟨such⟩ remainders
> for two divisors is the product of the number of days ⟨elapsed in the
> *Kaliyuga*⟩ and the reduced orbit of sky, therefore, having divided
> ⟨it⟩ by their own divisors, the quotient is the passed revolutions of
> the sun and the moon.

In other words, since

$$N = A_y \times K_1 \times \lambda_1 = A_y \times K_2 \times \lambda_2,$$

then

$$\lambda_1 = \frac{N}{A_y \times K_1} \lambda_2 = \frac{N}{A_y \times K_2}$$

And, as Bhāskara adds:

> *asminn eva dvicchedagre apavartitakhakakṣyayā vibhakte labdham*
> *ahargaṇaḥ*
> When that which has ⟨such⟩ remainders is divided by the reduced
> orbit of the sky, the quotient is the number of days ⟨elapsed in the
> *Kaliyuga*⟩

In other words,

$$x = \frac{N}{K}.$$

c   Case with two planets and the residue of minutes in respect to the planet's orbits   The problem is the same as before, only instead of the residue of revolutions in terms of the planet's orbits the residue of minutes Rk_{L_1} and Rk_{L_2} are given.

Two procedures are given to find the integral number of revolutions, signs, degrees and minutes crossed by both planets since the beginning of the *Kaliyuga*:

Procedure 1 If y (resp. z) is the integral number of revolutions, signs, degrees and minutes crossed, in terms of revolutions by the first planet (resp. the second planet), then the problem may be formalized as

$$z \times 21600 = \frac{A_y K_1 \times 21600y - (Rk_{M_2} - Rk_{M_1})}{A_y K_2}.$$

It can be solved by any of the two methods used for this type of problem (a pulverizer without remainder or a *sthirakuṭṭāka*).

Procedure 2 In this case the residue of degrees in terms of the planet's orbit (Rk_B) is found by solving the problem

$$y_B = \frac{60 \times x_B - Rk_L}{A_y \times K},$$

where x_B is the residue of degrees in terms of the planet's orbit, and y_B the integral number of degrees crossed by that planet.

Then the residue of signs in terms of the planet's orbit (Rk_R) is found by solving the following problem:

$$y_R = \frac{30 \times x_R - Rk_B}{A_y \times K},$$

where x_R is the residue of signs in terms of the planet's orbit, and y_R the integral number of signs crossed by that planet.

From this the residue of revolutions in terms of the planet's orbit (Rk_M) is found by solving the following problem:

$$y_M = \frac{30 \times x_M - Rk_R}{A_y \times K},$$

where x_M is the residue of revolutions in terms of the planets' orbit, and y_M the integral number of revolutions crossed by that planet.

d Case with more than two planets This combines the above described procedures, with the case of the problems where what is sought is an integer N having given remainders for n different divisors.

For a first two couple of planets, g_1 and g_2, N_1 is found as described above, for the couple of divisors and remainders $(A_y K_1, Rk_{M_1}; A_y K_2, Rk_{M_2})$, if the residue of revolutions in terms of the planet's orbits is given. Then for a third planet, g_3, the same procedure is applied to the couple $(A_y^2 K_1 K_2, N_1; A_y K_3, Rk_{M_3})$. And so forth.

Appendix: Some elements of Indian astronomy

1 Generalities

The sky is considered as a sphere (*gola*) whose radius is 3438 minutes (*kalās*)[128], with the earth at its center. Stars are fixed on the sphere, which is thus called *bhagola*, "sphere of the asterisms/stars". We will call it here the Celestial sphere. Tradition states that the earth does not move, and that the Celestial sphere turns daily around the line going from the North pole (P) to the South pole (P') called the Celestial axis. Āryabhaṭa however considered that the earth rotated from West to East, and therefore that the movement of the Celestial sphere was only apparent. Because of the violent reactions such a statement provoked, later commentators changed the verse in order for it to mean exactly the contrary[129]. The planets, among which the sun and the moon, revolve in the space between the earth and the Celestial sphere. The Celestial Equator (*viṣuvat*) is defined as the great circle (i.e. a circle belonging to the sphere and having the earth for center) perpendicular to the Celestial axis.

Let us imagine an observer (O) on earth. Since the earth and thus the point where the observer stands is very small compared to the radius of the Celestial sphere, both are collected together. Apart from the Celestial axis and Equator, in the following representations, all the other planes and lines will be defined according to this observer.

The imaginary vertical line, which through the observer's feet extends itself to two points on the surface of the sphere, defines respectively the zenith (Z, *nata*), which is the point above, and the Nadir (Z'), which is the point below. This is illustrated in Figure 56.

---

[128]The reason why a circular measuring unit is used here remains mysterious to me.

[129]See for instance [Sharma&Shukla 1976; Intro, p.xxix; p. 8; p. 119-120], [Yano 1980], and [Bhattacharya 1991]

Figure 56: The Celestial sphere
$NESW$ is the Horizon for the observer in O;
$ZPZ'P'$ is the Celestial Meridian;
$ZEZ'W$ is the prime vertical;
$WQEQ'$ is the Celestial Equator.

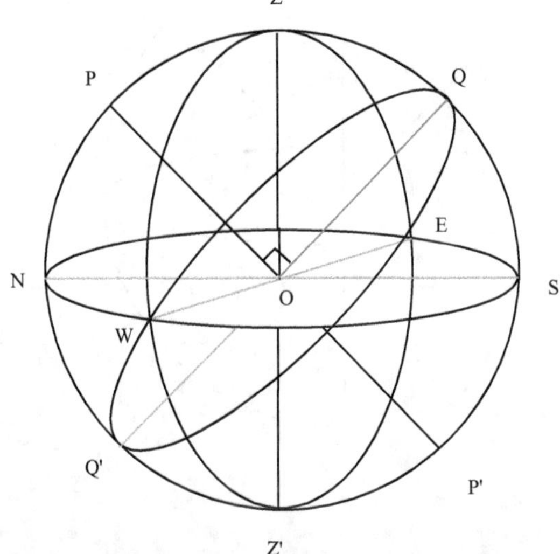

The great circle perpendicular to ZOZ' is called the Horizon. The plane it encloses is the plane of the observer. It intersects the Celestial Equator in two points called the East (E) and West (W).

The great circle which passes through the zenith, nadir and the poles is called the Celestial Meridian for this observer. It intersects the Horizon at the North (N) and South (S).

The great circle perpendicular to the Celestial Meridian, passing through the zenith and Nadir, and the the East (E) and the west (W) is called the prime vertical (*samamaṇḍala*).

2 Coordinates

The latitude of the observer, O, usually noted ϕ, is the angular distance between the Equator and the zenith (the arc ZQ as illustrated in Figure 57.)

The distance of the pole to the Horizon (the arc PN) is called the altitude of the pole. Because the angles ZOQ and PON are equal, the altitude of the pole and the latitude of the observer are equal. The co-latitude is $90° - \phi$ (as the arc QS).

Figure 57: Coordinates
ϕ is the latitude of the observer in O;
$90 - \phi$ is the co-latitude.

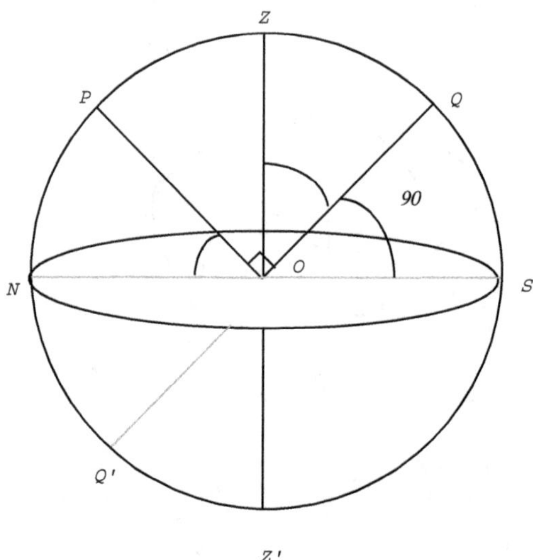

Let us now consider the orbit of the sun.

The path of the sun in the sky relatively to the stars, and to a fixed earth, when noted during a year, at a given time, in a given place, every day, draws an ellipse. This ellipse is in fact a mirror of the motion of the earth around the sun. The plane defined by this ellipse intersects the Celestial sphere in a great circle called the Ecliptic (*apamaṇḍala*). The Ecliptic intersects the Celestial Equator in two points γ and Ω. The angle of the sun with the Equator is constantly changing. In γ and Ω it is zero. The points where it is the greatest is called the obliquity of the Ecliptic (*paramāraprama*, lit. "greatest declination"). This is illustrated in Figure 58, page 189.

Today this angle, which is also that of the Ecliptic with the Equator, is roughly considered to be $23°7'$. γ is the point of the Equator through which the sun is considered to move from the southern hemisphere to the northern hemisphere. It is called the vernal equinox. Ω is the point on the Equator through which the sun is considered to move from the northern hemisphere to the southern hemisphere. It is called the autumnal equinox. The two points where the sun is at its greatest angular distance from the Celestial Equator are called the summer (Y) and Winter (M) solstice.

The Ecliptic represents the yearly path of the sun on the Celestial sphere. Daily, however, the sun is considered to have a motion parallel to that of the Equator,

Figure 58: Apparent motion of the sun in a year
γ is the vernal equinox;
Ω is the autumnal equinox;

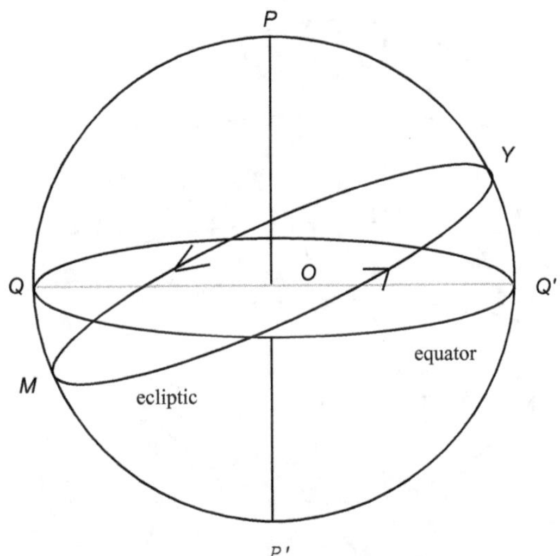

because of the rotation of the Celestial sphere around the axis of the poles. In fact, if we would represent the daily motions of the sun in a year, it would appear as a spiral made of roughly 365 spins parallel to the Equator. It would be a spiral because in 24 hours the sun slightly moves along the Ecliptic. During the vernal and autumnal equinox the apparent motion of the sun is on the Equator. The days are equal to the nights. The day of the winter solstice is the shortest of the year. The day of the summer solstice is the longest of the year. Whatever the day, at mid-day the sun is on the Celestial Meridian. This is illustrated in Figure 59, page 190.

Let's take any day of the year, and consider the sun at mid-day, as illustrated in Figure 60, page 190.

The straight line $SuSu'$ represents the orbit of the sun. At mid-day the sun is in Su. The angular distance between the zenith and the sun at Su (the arc ZSu) is called the zenith distance of the sun (z). The angular distance between the Horizon and the sun at Su is the altitude of the sun (a).

On an equinoctial day, the sun is on the Celestial Equator, as illustrated in Figure 61. At mid-day the sun is in Q. The zenith distance of the sun in Q is then the latitude (*akṣa*) of the observer. And its altitude becomes the co-latitude (*avalambaka*) of the observer.

These concepts are used in Bhāskara's commentary, when studying the astronom-

Figure 59: Daily and yearly apparent motions of the sun

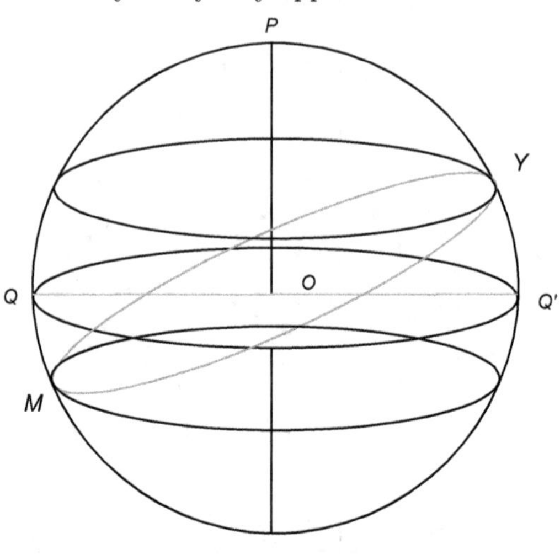

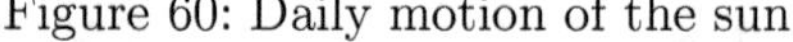

Figure 60: Daily motion of the sun

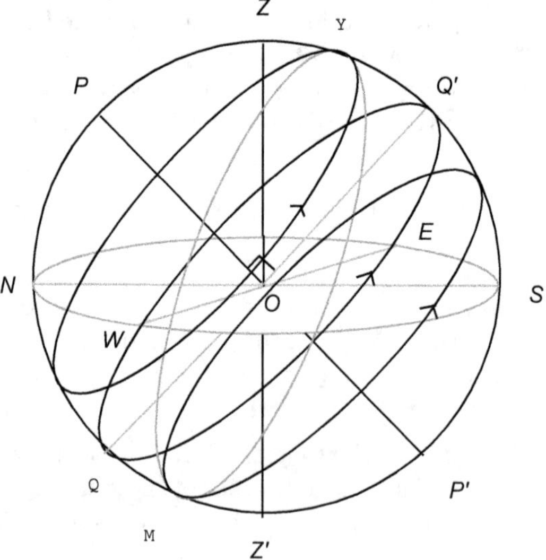

Figure 61: The sun on an equinoctial day

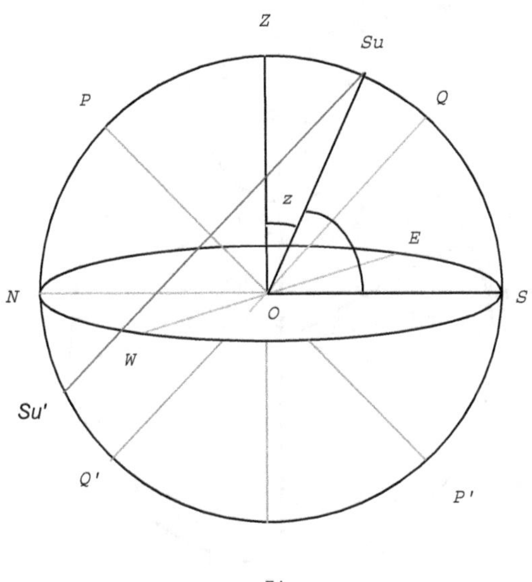

ical interpretation of the shadow cast by a gnomon, at mid-day (in BAB.2.14).

3 Movement of planets

One aspect of the Hindu planetary theory bearing traces of a Hellenistic influence concerns the description of the apparent motion of planets. These are rendered through an epicycle theory: the problem then being the constant discrepancy between the mean motions and the true ones. We will expose very briefly here some elements of Bhāskara's epicyclic theory. For a more detailed analysis see the explanations given in Chapter IV of [Shukla 1960].

A planet G (*graha*) has a mean circular motion, along a great circle of the Celestial sphere, the deferent, called in Bhāskara's commentary *vyāsārdhamaṇḍala* ("the circle ⟨of that⟩ semi-diameter"). Āryabhaṭa calls it *kakṣyāmaṇḍala* (Ab.3.18) "orbit's circle". Let O, the earth, be its center, and R, the radius of the celestial sphere, its radius. This is illustrated in Figure 62.

However, at a specific time of a specific day, the tabulated position of G is considered to be on a second smaller circle, the epicycle (*pratimaṇḍala*), which revolves in a direction opposite to the revolution described by the deferent. Although the point on the epicycle representing G at that time on that day is not yet the true position of G, it is considered a first, better approximation of it.

Figure 62: Orbit of a planet

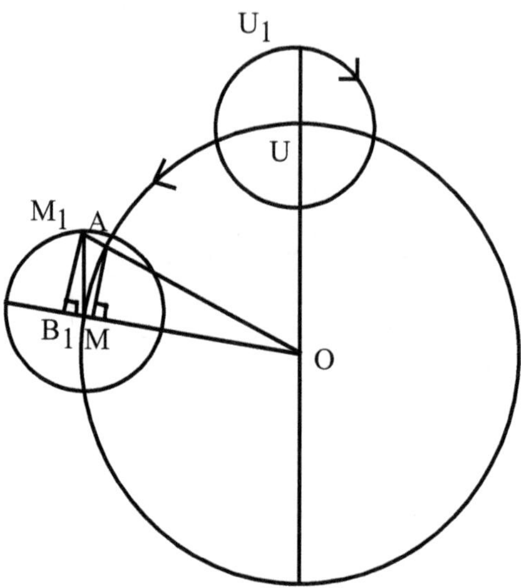

Let U_1 be the apogee (*ucca*) of G. Bhāskara defines in BAB.3.4ab [Sharma& Shukla 1976; p.179, line22-23], the *ucca* as follows:

> *yatra grahāḥ sūkṣmā lakṣayante (Shukla's readings)/labhyante (Mss. reading) karṇasya mahattvāt sa ākāśapradeśa uccasaṃjñitaḥ*

That we can understand as follows:

> A spot in the sky where a planet is perceived to be small because of the greatness of the hypotenuse (*karṇa*) is called *ucca* (high).

The apogee is the apparent remotest point of G along its orbit, and U is its mean position along its orbit. UU_1 serves as reference both for the radius of the epicycle at any time, and for the exact place on the epicycle where the tabulated position of G on the epicycle should be.

Let M be the mean position of G on its circular orbit on a given day at a given time. The arc UM represents the mean arc distance of G to its apogee at that given time, and is called the *bhujā*. Let M_1 be an approximation of the true position of G when its mean position is in M. M_1 is such that $MM_1 = UU_1$. This defines the epicycle. In his commentary on Ab.2.26-27.ab, Bhāskara does not consider the epicycle itself, but the circle having for radius OM_1: *tatkālotpannakarṇaviṣkambhārdhamaṇḍala* (the circle which has for semi-diameter the hypotenuse produced at that time).

Let A be the point of OM_1 that intersects with the mean orbit of G. Let B be a point of (MO) such that AB is perpendicular to (MO). Let B_1 be a point of (MO) such that M_1B_1 is perpendicular to (MO). Both AB and M_1B_1 are called the *bhujāphala* (the correction of the *bhujā*). OA is the radius of the orbit (*vyāsārdha*) and OM_1 is called the hypotenuse (*karṇa*).

Bhāskara states in BAB.2.26-27.ab that

$$\frac{AB}{OA} = \frac{B_1M_1}{OM_1},$$

and thus that AB is inversely proportionate to OM_1.

This section and the following give several supplementary remarks on the astronomical aspects of BAB.2.32-33.

4   Time cycles

Traditional Hinduism considers time as cyclical: there are four ages, called *yugas*, at the end of which the universe is destroyed and reborn again. The four *yugas*, in which the conditions of life increasingly deteriorates, are in due order: the *kṛtayuga*, the *tretāyuga*, the *dvāparayuga*, and the *kaliyuga* in which we presently live.

Ab.1.3-4 gives the numbers of revolutions of the sun, moon, earth etc. in a *yuga*, and the date of the beginning of the current *yuga*. Ab.3.5 defines solar years (*saṃvatsara*), lunar months and civil and sidereal days. A solar year is defined by the time taken by the sun, apparently, to make a full rotation around the earth. The number of solar revolutions, which gives the number of years, in a *yuga* is stated to be 4 320 000.

Traditional astronomy also distinguishes between civil days (*bhūdivasa/dina*, lit. terrestrial days) and celestial ones (*nakṣatradivasa*). A celestial day corresponds to one apparent rotation of the celestial sphere from East to West. A civil day corresponds to the daily apparent rotation of the sun around the earth: since the sun every day slides slightly on the ecliptic there is a discrepancy between celestial and civil days.

The civil days are defined in Ab.3.5: "The conjunctions of the sun and the earth are (civil) days"[130]. The computation of the number of conjunctions in a *yuga* is defined in Ab.3.3ab: "The difference between the revolution-numbers of any two planets is the number of conjunctions of those planets in a *yuga*." [131] The "revolution-number" (*bhagaṇa*) of a planet is the number of revolutions of a planet in a *yuga*: these are constant and given in Ab.3-4. The number of terrestrial revolutions in a *yuga* is given by Āryabhaṭa in Ab.1.3: 1582237500. So that the number of civil days in a *yuga* ($A_y{}^{132}$) is equal to the number of revolutions

---

[130] [Sharma&Shukla 1976; p. 91]

[131] op. cit., p.86.

[132] This corresponds to the notations we have adopted in our supplement for BAB.2.32-33.

of the sun in a *yuga* minus the number of revolutions of the earth in a *yuga*: 1582237500 − 4320000 = 1577917500. Therefore $A_y = 1577917500$.

This value is important when evaluating the number of days elapsed in the *Kaliyuga*, when the longitude of a given planet is known. This is one of the astronomical problems solved by a pulverizer computation, as described by Bhāskara in BAB.2.32-33.

5   Orbits and non-integral residues of revolutions

The mean orbit (*kakṣyā*) of a planet, as we have seen above, is considered to be a circle (*kakṣyāvṛtta*). It represents the apparent motion of a planet, around the earth, on the Celestial sphere. One movement of the planet along its orbit is called a revolution (*maṇḍala*). A revolution is divided into twelve equal signs (*rāśi*). A revolution is also divided into three hundred and sixty degrees (*bhāga*), so that there are thirty degrees per sign. A degree is divided into sixty minutes (*liptā*), a minute into sixty seconds (*vikalā*)[133]. This is summed up in Table 11.

Table 11: The different subdivisions of a revolution

Sanskrit	English	Respective Amounts				
		Rev	Signs	Deg	Min	Seconds
maṇḍala	Revolution	1				
rāśi	Sign	12	1			
bhāga	Degree	360	30	1		
liptā	Minute	216000	300	60	1	
vikalā	Second	1296000	18000	3600	60	1

At the beginning and at the end of a *yuga*, all planets are in conjunction. It is assumed that, along their respective orbits, all the planets cross the same distance in a *yuga*. This is stated in Ab.3.12 (op. cit. p. 100). The distance described by any planet in a *yuga* gives the "circumference of the sky"[134]. In verse 6 of the *Gītikāpāda*, Āryabhaṭa gives the following rule (given here with the non-literal translation by K.S. Shukla and K.V. Sharma op. cit., p.13) to compute the length in *yojanas* of the orbit of any planet:

Ab.1.6.

khayugāṃśe grahajavo

> The circumference of the sky divided by the revolutions of a planet in
> a *yuga* gives (the length of) the orbit on which the planet moves.

---

[133]These subdivisions, of course, recover those that divide a circle in mathematics. See the Section of the Glossary on time units.

[134]op. cit., p. 14

From this verse of the *Āryabhaṭīya* we also indirectly know that the circumference
of the sky in *yojanas* is: 12474720576000 *yojanas*. The orbit of the moon, according
to the value given in Ab.1.3, is

$$\frac{12474720576000}{57753336} = 216000 \ yojanas.$$

And the orbit of the sun is

$$\frac{12474720576000}{4320000} = 2887666, 8.$$

In the *Mahābhāskarīya*, the following verse gives a rule to find the mean longitude
of a planet [135]:

> Mbh.i.20
> *ambaroruparidhir vibhājito bhūdinair divasayojanāni taiḥ|*
> *saṅguṇayya divasān athā haret kakṣyayā bhagaṇarāśayaḥ svayā||*
>
> Divide the (*yojanas* of the) circumference of the sky by the number of
> civil days (in a *yuga*): the result is the number of *yojanas* traversed (by
> a planet) per day. By those (*yojanas*) multiply the *ahargaṇa* and then
> divide (the product) by the length (in *yojanas*) of the own orbit of the
> planet. From that are obtained the revolutions, signs, etc. (of the mean
> longitude of the planet).

The *ahargaṇa*, is the number of days elapsed in the *Kaliyuga* at that time. If x is
the *ahargaṇa*, since we know that the number of civil days in a yuga is 1577917500,
then, for example, the mean longitude of the sun (λ_S) is

$$\lambda_S = \frac{12474720576000x}{1577917500 \times 2887666, 8}.$$

We can recognize here the type of problem solved by a pulverizer without remainder. Such problems are seen in Examples 24-26 of BAB.2.32-33. Note that there
would be an obvious simplification here, that does not seem to be carried out in
the resolution of these examples:

$$\lambda_S = \frac{12474720576000x}{1577917500} \times \frac{4320000}{12474720576000} = \frac{4320000x}{1577917500} = \frac{576x}{210389}.$$

---

[135][Shukla 1960; Skt, p. 4; Eng, p.15]

Glossary

1 General

The words are given in the Sanskrit order. Double quotes indicate the technical translation chosen, as opposed to the literal translation of a word or expression. Are noted as synonyms, those that are given as such by Bhāskara[136].

A

Akṣa Latitude. *Akṣajyā* The Rsine of the latitude.

Akṣepa Non-additive. Said of two *karaṇīs* that cannot be summed.

Agra Remainder. In one instance of far-fetched interpretation (BAB.2.32-33), Bhāskara understands this word used in Āryabhaṭa's verse as meaning "a number".

Adhikāgrabhāgahāra or adhikāgraccheda Technical term of the *kuṭṭakāra* procedure. It is "the divisor of the greater remainder" in a pulverizer with remainder (*sāgrakuṭṭakāra*) procedure. It is "the divisor which is a large number" in the pulverizer without remainder (*niragrakuṭṭaka*) procedure.

Anuloma Same direction. Direct. *Anulomagati* is a direct motion, as opposed to *vilomagati*, a retrograde motion. *Anulomacārin* has the same meaning. *Anulomavivara* is the distance of ⟨two bodies moving in⟩ the same direction.

Anta Last term of a series.

Antara Distance, difference. *Deśāntara*, lit. difference of spots, is the "longitude". *Sthānāntara* is a different place. In common Sanskrit it means particular, as in *upāyāntara* (a particular method) or different, as in *ābhādhāntara*: the different sections (of the base).

Antarāla Space between. An interval.

---

[136]Please see in the section "Conventions of translations" in *Introducing the Translation*, the paragraph on synonyms, for a short discussion of this topic.

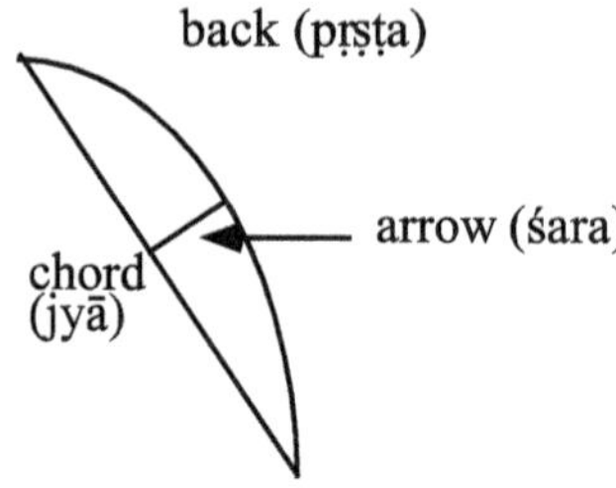

Figure 63: A bow-field

Antya Last.

Apacaya Decrease; Subtractive (quantity), subtrahend.

Apanayed One should subtract.

Apavartita, apavartya Reduced (by a common factor).

Aparvartana Division. Reducer (as one who does the action described as *apavartita*). Given as a synonym of *bhāga* (division, part) in BAB.2.4.

apa-VṚT To reduce (by a common factor), to divide.

Abhyasta Multiplied.

Abhyāsa Product. The product *of* two or more quantities, as opposed to the multiplication of a quantity *by* another.

Aṃśa Part. numerator of a fraction. A fraction. When a fractional number is stated, the denominator is marked with *aṃśa*. Also used as a substitute for *bhāga* with the meaning of "degree".

Ardha Half. Increase in commercial problems.

Ardhita Halved.

Avagāhya, avagāha Penetration. Lit. "having plunged". Segment of the diameter of a circle. Also used for the arrow (*śara*), of a bow-field (illustrated in Figure 63)

Avayava Part.

Avarga see *varga*.

Avalambaka Perpendicular. Plumb-line. Rsine of the co-latitude. The Rsine of the co-latitude is proportional, on an equinoctial day, to the perpendicular formed by the body of a gnomon.

Avasāna Distance. Literally it means a boundary. Only used in BAB.2.16 to refer to the distance between a gnomon and a source of light.

Aśeṣagaṇita Mathematics as a whole, i.e. mathematics seen as a global subject. See *gaṇita*.

Aśra or Aśri Side, edge. Used in the names of planes and solids.

A *caturaśrakṣetra* A quadrilateral field, and a *dvādaśāśri* "a twelve edged ⟨solid⟩", which is one of the names, here, for a cube. However in BAB.2.14. a *caturaśra* is used to qualify a solid – this may be another name for a cube, or that of a prism.

A *tryaśrakṣetra* A "trilateral field" and a *ṣaḍaśri* is "a six-edged solid", which is the name, here, of an equilateral pyramid with a triangular base. However in BAB.2.14, a *tryaśra* is used to qualify a solid, maybe a pyramid with a triangular base.

Asata Incorrect ⟨value⟩. Companion term of *sata* (correct ⟨value⟩).

Ahargaṇa Lit. group of days, is the number of days elapsed since a given epoch, usually the *Kaliyuga*.

Ā

Ācārya Master, teacher, learned one. It is often attached, as an honorific suffix, to the name of a person.

Ādi The first term of a series.

Ānayana To compute, computation. Mostly used in the introductory sentence, preceding the quotation of a verse of the *Āryabhaṭīya* about to be commented, which gives the aim of the procedure which will be treated.

Ābhādhā Technical term naming a segment of the base delimited by a perpendicular.

Āyata Elongated. length. *Āyatacaturaśrakṣetra*, lit. elongated quadrilateral field is always a rectangular field.

Āyāma Length. In a trapezium, it is one of the names of the height; length in a rectangle as opposed to *vistāra* which then means width.

Ārya This is the meter in which the three last quarters of the *Āryabhaṭīya*, including the *gaṇitapāda*, are written.

Ālekhya Lit. written, painted. A "drawing".

Āsanna Approximate, approximation. Lit. close to. Companion term of *sūkṣma*, accurate.

However, *sūkṣmasya āsanna* is the approximation of an accurate ⟨value⟩. *Vyāvahārikasya āsanna* is the approximation of a practical value. The first being of better quality than the latter.

Figure 64: A tusk-field

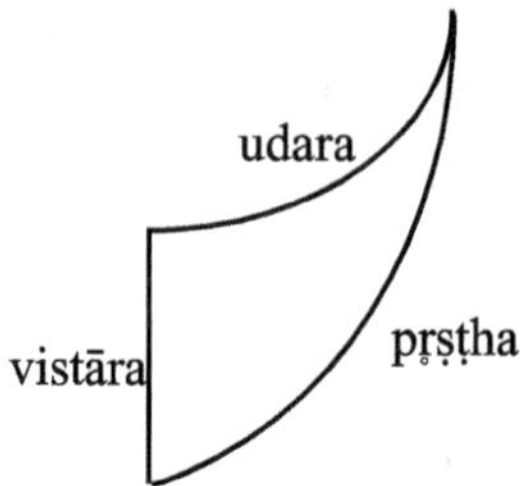

Āhniko bhogaḥ Daily passing. This is the name of the sum of the daily motions of two planets.

I

Icchā Desire.

> *Icchārāśi* The "desire quantity" in a Rule of Three. *Icchāphala* is the "fruit of the desire" in a Rule of Three.

Iṣṭa Desired. Sometimes close to the meaning of optional. In computation with series, *iṣṭa* is the desired number of terms.

U

Ucchrāya Height. Used when relating the geometrical cube to the square it is derived from, and when defining a triangular based pyramid.

Utkramajyā The Rversed sine, i.e Rsin. See the Annex to BAB.2.12.

Uttara The common difference in arithmetical series. Increase.

Udara Belly. Used to characterize one of the sides of a tusk-field, see Figure 64.

Uddeśaka Example.

Uddeśana Example.

Udvartanā Multiplication. Given as a synonym of *saṃvarga* in BAB.2.3ab.

Upacaya Increase. Additive (quantity).

Upaciti Lit. accumulation. Is the name of the series of (the progressive sum of) natural numbers.

Upapatti Proof. Opposed to tradition (*āgama*) in BAB.2.10.

Uparirāśi See *rāśi*.

Upalakṣita Characterized.

Figure 65: Right-angled triangle in a *śṛṅgāta* field.

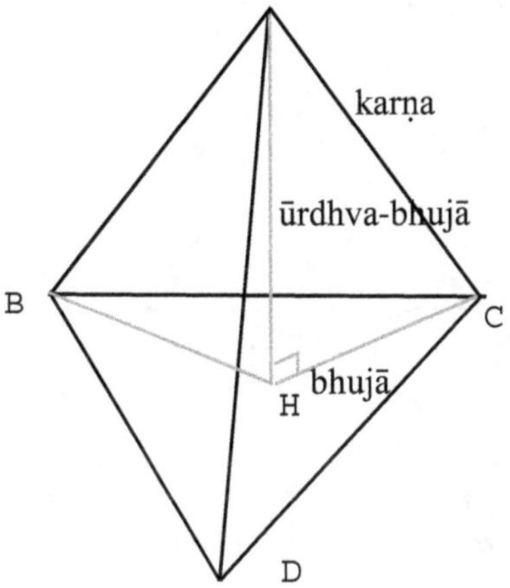

Upāya Method.

Ū

Ūna Decreased. Subtractive ⟨quantity⟩.

Ūnāgraccheda or ūnāgrabhāgahāra Is "the divisor for the smaller remainder" in a pulverizer with remainder and "the divisor which is a small number" in a pulverizer without remainder procedure.

Ūrdhvabhujā Upward side. Used for the perpendicular issued from one vertex on to the triangular base in a *śṛṅgātaka* field, as illustrated in Figure 65.

Ṛ

Ṛkṣa Sign. 1/12th of the circumference of a circle.

Ṛju Vertical.

Ṛjuta Verticality. *Ṛjusthiti* is a steady vertical.

Ṛṇa Debt. When opposed to *dhana* (wealth) it is a "subtractive ⟨quantity⟩".

E

Ekatra kṛtvā Summed. lit. having made in one place; this may refer to the fact that the two summed quantities were erased from the working surface, and replaced by one quantity, their sum, that occupied thereafter only "one place" on the working surface.

Figure 66: Right-angled triangle

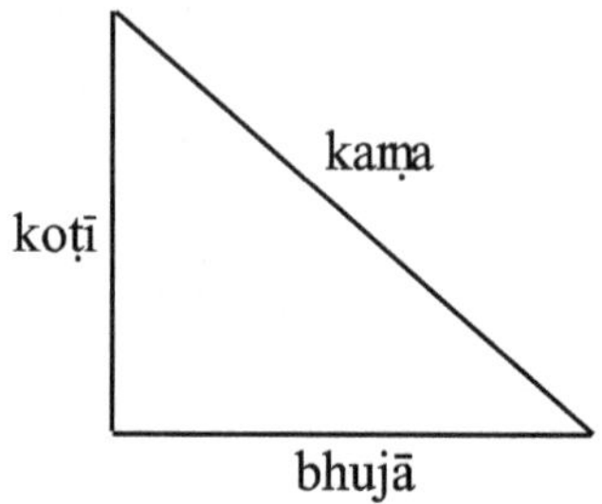

Ekī bhavā, ekī kṛtya Sum. Lit. the state of becoming one, having made into one; this may refer to the fact that the two summed quantities were erased from the working surface, and replaced by one quantity, their sum. See *ekatrakṛtvā*.

Ka

Kakṣyā Orbit of a planet.

Karaṇa Procedure. Name given to the part of an example which exposes its resolution.

Pratilomakaraṇa Is a reversed procedure.

Karaṇika Which belongs to *karaṇīs*, which measures the *karaṇī* (of a given quantity).

Karaṇika Is derived from the word *karaṇī*, to which the suffix *-ka* is added, followed by a diminution of the long ī.

Karaṇī Usually considered as a "surd", the expression "the *karaṇīs* of *a*" may be translated as meaning: "that whose square is *a*", or $\sqrt{a}$. However, it seems to be a geometrical concept. It may be a specific way of considering the square of the measure of a geometrical object (see the section 1 of part I). It is given as a synonym of *varga* in BAB.2.3ab.

karaṇīparikarman The geometrical operation of constructing the square having the hypotenuse for side: its area is equal to the sum of the two other sides of a right-angle triangle, as well as the numerical squaring of the length of the hypotenuse as the sum of the squares of the two other sides.

Karidantakṣetra A (two dimensional) tusk-field. see Figure 64, page 200.

Karkaṭa, karkaṭaka Lit. a crab; it is the name of " a pair of compasses".

Karṇa hypotenuse. Diagonal. In customary Sanskrit it is an "ear".
Karṇa is used in the traditional enumeration of the sides of a right-angle triangle: *karṇabhujākoṭi*. See Figure 66.

We will use the literal translation when it is used to describe the side of a field or a solid, where no right-angle triangle is immediately involved. But usually it names a segment of a geometrical figure, in which Ab.2.17 (i.e. the "Pythagoras Theorem") may be applied; when this is the case, it becomes then the hypotenuse of a right-angle triangle, and we have translated it accordingly. See for instance Figure 65, page 201.

In any triangle, the sides for a given base are also called *karṇa*, which means "ears". These may also be named by synonyms of this term as *śravaṇa* and so forth.

Karman Computation, operation.

> *Gaṇitakarman* A mathematical operation. *viparīta-*, *pratiloma-* and *vilomakarman* mean a reversed operation.

Kārikā Verse.

Kāla Time. *Kālakrīyā*," time reckoning" is the third chapter of the *Āryabhaṭīya*.

> *yogakāla* The meeting time (of two moving bodies).

Kāṣṭha Unit arc. This is a terminology particular to Bhāskara. It glosses Āryabhaṭa's use of *capa* in Ab.2.11 but can be found in the *Mahābhāskarīya* as well[137].

Kuṭṭākāra or Kuṭṭāka Pulverizer. Name of the procedure described in verses 32-33 of the Chapter on mathematics of the *Āryabhaṭīya*.

> *Sāgrakuṭṭākāra* A pulverizer with remainder. *Niragrakuṭṭāka* is a pulverizer without remainder.

> *Velākuṭṭākāra* The time pulverizer.

Kṛti Square. Given as a synonym of *varga* in BAB.2.3ab.

Kendra Center.

Koṭī or koṭi The upright-side; see Figure 66.

> It is usually one of the sides of a right-angle triangle, the other one is called *bhujā*, and the hypotenuse *karṇa*. See Figure 66. This word is also used to name the vertical edge of a gnomon.

Krama Method.

Kriyyā Method.

Kṣaya Decrease.

---

[137]See Shukla's remark in [Shukla 1976; Intro, p.xlii]

Kṣetra Field, and by extension a geometrical figure. It sometimes refers to the *surface* delimited by a number of sides or a line. It sometimes refers only to the *set of lines and inner segments* that draw the field, and not to the delimited surface.

Kṣetragaṇita The mathematics of fields or computations with fields.

Kṣepa Additive ⟨quantity⟩.

Ga

Gaccha The number of terms in a series. In one instance, BAB.2.20, it is also interpreted as a term of the series. This would be rather Āryabhaṭa's understanding of the word, rather than Bhāskara's. In BAB.2.29 it is a term of a set: *pada* and *paryavasāna* are given as synonyms of this word.

Gaṇaka Mathematician? A literal translation would be computer (in the sense of someone who computes), we have translated it by "calculator".

Gaṇita Mathematics. computation. By extension *gaṇita* sometimes names the result of any computation, and therefore means sometimes: area, sum, quantity.

Aśeṣagaṇita Lit. mathematics without remainder, is "mathematics as a whole" which englobes both *samānyagaṇita*, general mathematics, and its counterpart, *viśeṣagaṇita*, "specific mathematics".

Gaṇitakarman A mathematical operation.

Średdhīgaṇita The sum of a series.

Laukikagaṇita Is wordly computations.

Gata Lit. gone, "exponention"; i.e. the raising to any power of a quantity. The word with this technical meaning is only used in BAB.2 introduction.

A *dvigata*, a double-*gata*, is a square (*varga*); a *trigata* is a cube (*ghana*). By the same token, *gatasya mūla* or *gatamūla*, lit. the root of a *gata*, is a root extraction from any power.

Gāthā Synonym of *ārya* as a name of a verse-meter.

Guṇa Multiplier. Occasionally translated as "times".

Guṇakāra Multiplier.

Guṇanā Multiplication (of two different quantities, counterpart of the term *gata*); however it is given as a synonym of *saṃvarga* in BAB.2.3ab.

Guṇita Multiplied. This word is given as a synonym of *hata* in BAB.2.7.ab.

Guṇya Multiplicand.

Gulikā Bead. Name of the coefficient of the unknown quantity in first order equations.

Gūha Sign. 1/12th of the circumference of a circle.

Gola, golaka Sphere.

Golapāda The name of the fourth chapter of the *Āryabhaṭīya*.

Ghanagola A circular solid.

Graha Planet.

Grahacāra The "motion of planets". *Grahagaṇita* is the "mathematics of planets/planetary computations".

Grāsa Lit. mouthful. Segment of the diameter of two intersecting circles. Name of the part of the sun eclipsed by the moon, or of the part of the moon eclipsed by the shadow of the earth (i.e. the part of the moon eaten by *Rāhu*).

Gha

Ghana Cube. Solid. A cube ⟨place⟩, i.e. in the decimal place-value notation it is a place whose power of ten is a cube. Conversely, a non-cube ⟨place⟩ is a place whose power of ten is not a cube.

Ghanaphala The volume. *Ghanamūla* is the cube-root.

Ghanagola A circular solid.

Citighana A solid ⟨made of⟩ a pile. This is the name used by Āryabaṭa for the series of the progressive sums of natural numbers (i.e. the sum of $1, 1 + 2, 1 + 2 + 3, \cdots, 1 + 2 + 3 + \cdots + i, \ldots$).

Ghāta Multiplication. Given as a synonym of *saṃvarga* in BAB.2.3ab.

Ghna Multiplier.

Ca

Cakra A revolution. In customary Sanskrit it is a circle.

Caturaśrakṣetra Quadrilateral field. Sometimes the term field (*kṣetra*) is omitted in which case we translate the compound as 'quadrilateral'. Means literally: " a field with four sides".

Cāpa Unit arc.

Citi Pile. Used in the geometrical description of series.

Cha

Chāya Shadow. Rsine of the zenith distance. It is the name of a specific field of mathematics, related to computations using the data given by a gnomon. It is the length of the midday shadow cast by a gnomon. It is proportionate to the Rsine of zenith distance which thus sometimes bears the same name.

Figure 67: The diagram in BAB.2.11

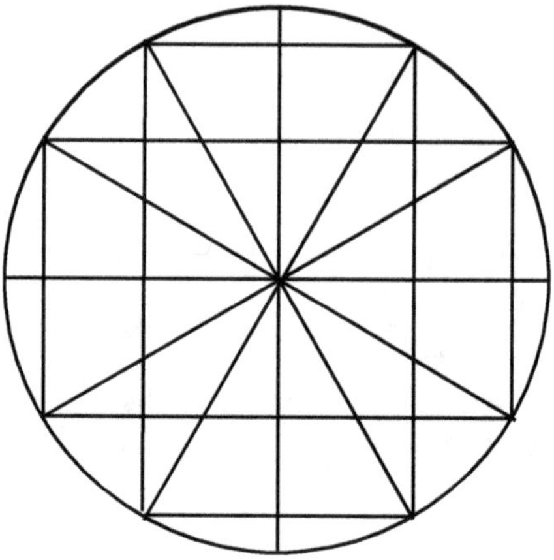

Chindyāt One should divide.

Cheda Part. Denominator of a fraction. divisor. In BAB.2.12 once used as meaning "partial (half-chord)".

Adhikāgraccheda The divisor of the greater remainder.

Ūnāgraccheda The divisor of the smaller remainder.

Chedyaka A diagram. In this commentary the word is only used in reference to a specific diagram, whose construction is described in BAB.2.11., with which the measure of half-chords (*ardhajyā*) or Rsinuses (R times the sinus) is derived. See Figure 67.

Ja

Jīvā A chord.

Jyā Chord.

Ardhajyā A half-chord. Half the chord subtending the arc 2α $(\frac{crd(2\alpha)}{2})$ is called the half-chord of α. This is what we call *Rsinα*, see Figure 68 and the Annex to BAB.2.11.

By extension *jyā* is sometimes the half-chord.

Jyotpatti A production of ⟨half⟩-chords.

Jyāvibhāga A partition of chords. In BAB.2.11, this refers to the subdivision of the perimeter of the circle into equal arcs and to the interior fields drawn inside the circle, as illustrated in Figure 67. In BAB.2.12, this refers, along with other expressions as "*khaṇḍitaṃ ... ardham*" (the expression used by

Figure 68: Chord and half-chord

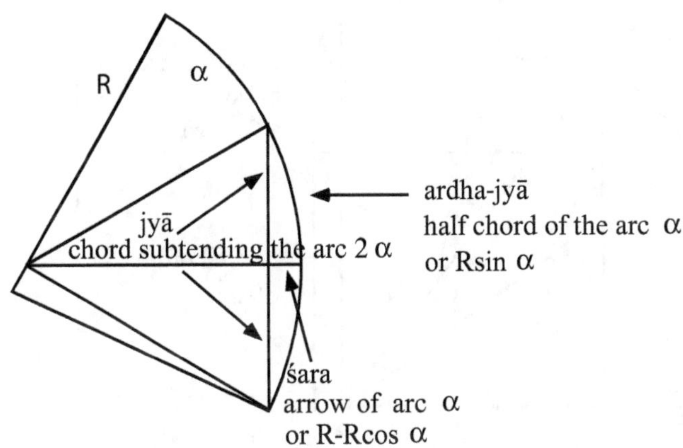

Āryabhaṭa in Ab.2.12) and *"chinnaṃ ... ardhaṃ"* (the expression used by Bhāskara), both meaning "sectioned half⟨-chord⟩", to the difference of two successive half-chords. The difference of two half-chords appears as a segment of the biggest half-chord. See Figure 69.

Akṣajyā The Rsine of the latitude. *Natajyā* is the Rsine of the zenith distance.

Ta

Tatparās Seconds.

Tithi Lunar day.

Tulya Equal.

Trairāśika Rule of Three.

Da

Dalita Halved.

Dairghya Length. Given as a synonym of *āyāma* in BAB.2.8.

Dik (Cardinal) direction. Also used in a figurative sense.

Dina Day.

> *Dinarāśi*, lit. the amount of days, is the number of days elapsed in the *Kaliyuga*. *Dinagana*, lit. the group of days, has the same meaning.

Divicara Planet. lit. roaming the sky.

Figure 69: The difference of two half-chords

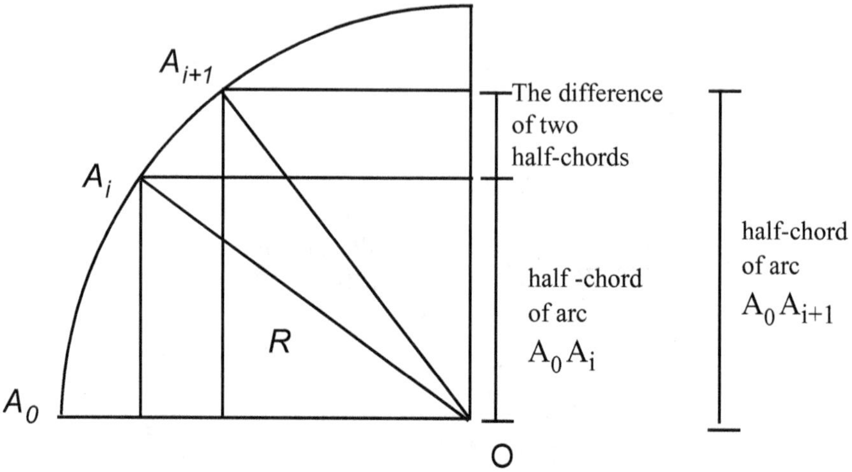

Dravya Object. Sum.

Dvicchedāgra ⟨A quantity that has such⟩ remainders for two divisors. Technical term denoting the number to be found in a "pulverizer with remainder" process.

Dha

Dhatrī Lit. earth, the "base" of a triangle, or the "earth" in a trapezium (the "earth" here is the base of the trapezium, but we have kept the literal translation here in order to distinguish it from its segments which can be the "base" of a triangle).

Dhana Lit. wealth. value, especially the value of the term of a series, i.e. the sum of the terms of a finit sequence. Amount. With the meaning of wealth as opposed to the word *ṛṇa* (debt) it is an additive quantity.

Madhyadhana The mean value, i.e. the mean sum of the terms of the sequence. *Sarvadhana* is the whole value, i.e. the sum of all the terms of a sequence. *Padadhana* is the value of the terms, which ambiguously may refer to the terms of the sequence or to its corresponding series.

Dhanuḥkṣetra Bow-field. It is made of an arc of a circle (called "the back" *pṛṣṭha*), the chord that subtends it (*jyā*) and an arrow (*śara*). It is illustrated in Figure 63, page 198.

Na

Naṭa Zenith distance.

> *Naṭajyā* The Rsine of the zenith distance.

Nāḍī Time unit equal to half a *muhūrta*, or 24 minutes.

Nirapavartita Reduced. See *apavartita*.

Niravaśeṣa Without remainder; without exception.

> *Niravaśeṣagaṇita* Is "mathematics as a whole", which englobes both *samānya-gaṇita* (general mathematics), and its counterpart, *viśeṣagaṇita*, (special(ised) / specific mathematics).

Nīyamāna Computing.

Nyāya Rule. Method. Logic?

Pa

Pada Term of a sequence, a series or of a set. Given as a synonym of *gaccha* in BAB.2.29. In Āryabhaṭa's understanding it would be the number of terms of a sequence. For Bhāskara however, its meaning is restricted to the meanings given as entries. Name given to the successive remainders that are placed, in the mutual division of the pulverizer (*kuṭṭākāra*) procedure.

> *Padapramāṇa* The number of terms in a series.

Paṇavakṣetra A drum-field. See illustrations in BAB.2.9.ab.

Parikarman Operation.

Parikalpaniyā Calculation.

Pariṇāha Circumference. Given in BAB.2.9.cd as a synonym of *paridhi*.

Paridhi Circumference, given in BAB.2.7ab and BAB.2.10 as a synonym of *pariṇāha*.

Parilekha The out-line ⟨of a circle⟩, i.e. the line that draws the circumference.

Parihāra Refutation.

Paryavasānam Term of a set. Given as a synonym of *gaccha* in BAB.2.29.

Pārśva Lit. a flank, it has the technical meaning of "side". In Āryabhaṭa's understanding it may be any side. In Bhāskara's understanding it may be restricted to orthogonal sides. It is however given by Bhāskara as a synonym of *bhujā* in BAB.2.6.ab.

> *Pārśvatā* "Sideness", maybe an expression meaning orthogonality.

Piṇḍita Added. This term is used by Āryabhaṭa rather than by Bhāskara.

Pṛṣṭha Back. Name of one of the sides of a tusk-field, see Figure 64, page 200 and of the arc of a bow-field, see Figure 63, page 198. This may be a general term for anything curved.

Prakriyā Calculation. In grammatical Sanskrit it means a derivation, i.e. what is done step by step.

Prakṣepa Sum. In commercial problems as the original sum invested by each member in a commercial transaction, so that it is sometimes translated as "investment".

Pratiloma Reversed.

> *Pratilomakaraṇa* is a reversed procedure. *pratilomakarman* is a reversed operation.

Pratyayakaraṇa Lit. a conviction-procedure, a "verification".

Pramāṇa Size, amount.

> *Pramāṇarāśi* The "measure-quantity" in a Rule of Three.

Pha

Phala Fruit; result. Thus the "interest" in commercial problems.

> *Kṣetraphala* The area. *Ghanaphala* is the volume. By extension, in a geometrical context, *phala* alone has sometimes been translated by area or volume. In a specific part of BAB.2.3cd *phala* is used as meaning 'surface', although this understanding can generally be attributed to the word *kṣetra* (field).

> *Phalarāśi* The " fruit quantity" in a Rule of Three.

> *Mūlaphala* The interest on the capital.

Ba

Bāhu Its usual meaning is arm or forearm, as a synonym of *bhuja* (given as such in BAB.2.6.ab), it is translated as "side".

Bīja Seed.

Brahma A pair of compasses. Terminology used by Āryabhaṭa.

Bha

Bhakta Divided.

Bhagaṇa Revolution.

Bhavana Zodiacal sign.

Bhāga Part; division. Degree, the 60th part of a circle or revolution in an astronomical context.

This word is derived from the verbal root *Bhaj-*, to share, distribute, which has the technical meaning "to divide". *Bhāgahṛtvā*, lit. when one has removed a part, means "when one has divided". *Bhāgalabdha* is what is obtained from the division or "the quotient of the division".

When expressing in words the fraction $\frac{a}{b}$, *bhāga* may be affixed to the denominator (b), thus meaning a out of b parts. It may also be affixed to the numerator (a), thus meaning a parts of b.

Śuddhaṃ bhāgaṃ, Lit. a pure division is "an exact division" that is it has no remainder.

Bhāgaśeṣa "The residue of degrees" , i.e the non-integer part of the number of degrees crossed by a planet since the beginning of the *Kaliyuga*.

Bhāgahāra Divisor. lit. removing a part.

Adhikāgrabhāgahāra The divisor of the greater remainder. A technical term of the *kuṭṭakāra* operation/procedure.

Bhajana Division. Given as a synonym of *bhāga* in BAB.2.4.

Bhāṣya Commentary. *Āryabhaṭīyabhāṣya* is the name of Bhāskara's commentary on Āryabhaṭa's work.

Bhinna Fraction, an integer increased or decreased by a fractional part, part.

Bhukti Daily motion.

Bhujā Side. In customary Sanskrit it is the corporeal arm. *Bhujā* can be any side of a field.

When considered in a *bahuvrīhi* compound, modified by *kṣetra*, it loses its *ā*: *tribhujakṣetra* is a trilateral field. Sometimes the word field (*kṣetra*) is omitted, the compound is then translated as "trilateral". *Caturbhujakṣetra* is a quadrilateral field.

Sometimes the meaning of *bhujā* is restricted to that of the base of a trilateral. *Bhujā* is one of the sides of the right-angle in a right-angle triangle, the other side is called *koṭī* and the hypotenuse *karṇa*. See Figure 66, page 202.

Bhujā In astronomy is the name for the mean arc distance of a planet at a given time, to its apogee. *Bhujāphala* "the correction of the *bhujā*" is a segment, which approximates the true position of a planet to its mean position at the same time. Please refer to the astronomical Appendix.

Bhū Lit. earth; the "base" of a triangle or the "earth" of a trapezium.

In the case of the trapezium, to distinguish it from its segments which may be the base of interior triangles, we have translated it as "earth". It is the companion term, in a trapezium, of *mukha* or *vadana*. See Figure 70

Figure 70: An isoceles trapezium

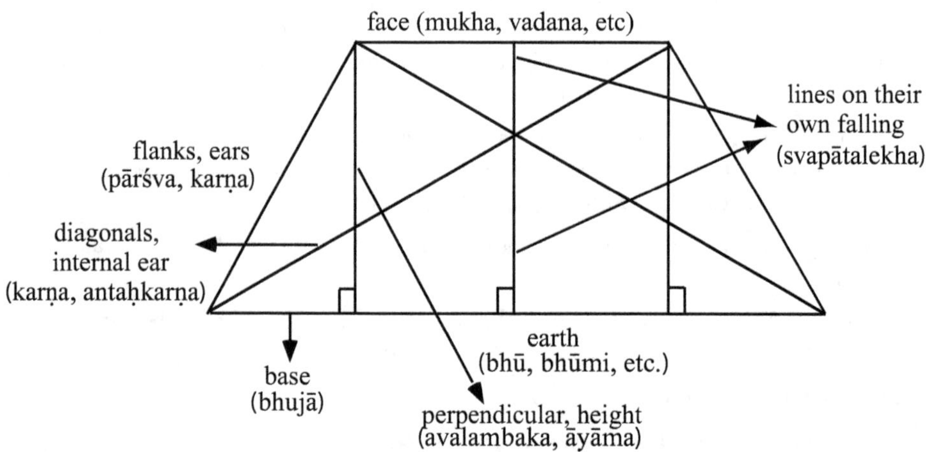

Bhūmi Lit. earth; the "base" of a triangle or the "earth" of a trapezium.

Bheda A part. Sometimes used figuratively, but also as the (fractional) part of a number.

Ma

Maṇḍala A circle. A revolution.

> *Maṇḍalaśeṣa* "The residue of revolutions", that is the non integer part of the number of revolutions performed by a planet since the begining of the *Kaliyuga*. This is illustrated in Figure 71.

Madhya Middle. Zenith. Mean.

> *Madhyadhanam* The mean value, i.e. the mean value of the sum of the terms of an arithmetical series.

Mahī Lit. earth, the "base" of a trilateral.

Miśrata Lit. mixture, "increased".

Mukha The face or mouth. Name of the side opposite to the earth in a trapezium. See Figure 70.

> It is also the name of the opening of a pair of compasses.

> The first term of a series.

Muhūrta Period of time equal to 48 minutes.

Figure 71: Residue of revolutions and residue of signs

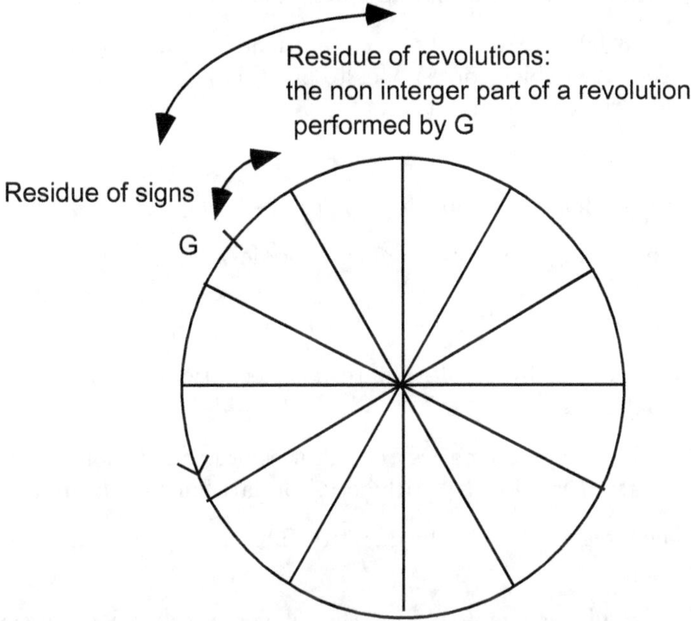

Mūla Root (in the common and mathematical sense). The "capital" in commercial problems.

> *Vargamūla* A square root. *Ghanamūla* is a cube root. *Gatamūla* is the root of an exponention, the fact of extracting a root. The latter compound is only used in BAB.2 introduction.

> In BAB.2.14 The word *mūla* is used to qualify the lower base of a gnomon.

> *mūlaphala* The interest on the capital.

Maurika Minute (as a unit used in longitudes).

Ya

Yāma Unit of time equal to 1/8th of a day or 3 hours.

Yāvakaraṇa Square. Given as a synonym of *varga* in BAB.2.3ab.

Yāvattāvat Lit. "as much as". Name of the the coefficient of the unknown quantity in first order equations. Used only by Bhāskara.

> *Yāvattāvatpramāṇa* The "value of the *yāvattāvat*, that which is unknown.

Yukta Increased. summed.

Yuktyā Adverb meaning "cleverly". The word *yukti*, with the meaning "reasoning", has an important posterity in Sanskrit mathematical texts.

Yuga A period of 4320000 years. There are traditionally four *yugas*, the last one being the *Kaliyuga* (which corresponds to our time) after which the earth is destroyed, and the cycle starts again.

Yuta Increased. Summed.

Yoga Sum. Meeting point (of two moving bodies).

yogakāla The meeting time (of two moving bodies).

Ra

Rāśi A quantity. Traditionally, the 12th part of the ecliptic. It is the 12th part of a circle or 30 degrees.

Rāśigaṇita Is lit. the mathematics of quantities or computations with quantities; we have translated it as "arithmetic" or "arithmetical computations".

Rāśirūpa The integer ⟨part⟩ of the quantity. This expression is solely used in BAB.2.26-27.ab.

Uparirāśi The "higher quantity", i.e. the integer in a fraction increased or decreased by a part; the disposion $\begin{matrix} a \\ b \\ c \end{matrix}$ corresponding to $a + \frac{b}{c}$.

Bhāskara uses in BAB.2.9.cd the expression *rāśidvayakṣetra*, a two-*rāśi* field, which would be the name of an arc measuring 60 degrees.

Rāśiśeṣa is "the residue of signs" that is the non-integer part of the number of signs crossed by a planet since the beginning of the *Kaliyuga* (measured in degrees and minutes). This is illustrated in Figure 71.

Rūpa A unit. A digit (i.e numbers from 1 to 9). A whole number.

rāśirūpa The integer ⟨part⟩ of a fractionary quantity.

Rekha A line. Used in the drawing of a diagram.

La

Lakṣaṇasūtra A rule which is a characterization. A way of expressing an abstract or general rule.

Labdha What is obtained, the result, the quotient when connected with division (*bhāga*).

Lava Degree. 1/30th of the circumference of a circle.

Liptā or Liptika Minutes, the 60th part of a degree.

Liptāśeṣa The "residue of minutes", that is the non-integer part of the number of minutes crossed by a certain planet since the beginning of the *Kaliyuga*.

Lekha A line. Used in the drawing of a diagram.

Parilekha The out-line of a circle, i.e. the line that draws its circumference.

Laukikagaṇita Wordly computation.

Va

Vadana Face, the side opposite to the earth in a trapezium. See Figure 70, page 212.

Varga Square. The geometrical square as well as the square of a number, according to Āryabhaṭa. Practicaly, Bhāskara uses it for the square of a quantity. The square-place, i.e. a place in the decimal place-value notation whose power of ten is pair.

Vargakarman, Square-operation, may be the squaring of the length of a diagonal in a quadrilateral or the hypothenuse of a right-angle triangle (*karṇa*). See the discussion in the Annex of BAB.2.3.ab.

Vargagaṇita Square computation. The squaring of a digit in the procedure of extraction of a square-root.

Vargamūla A square root.

Avarga A non-square place. In the decimal place-value notation, it is a place whose power of ten is odd.

Vargaṇā Square. Given as a synonym of *varga* in BAB.2.3ab.

Vasudhā Earth, in a trapezium, that is the side opposite to the face. See Figure 70, page 212.

Vastu Subject, substance, object. Used to indicate the subjects of the treatise.

Vikalā A second, a unit used in giving longitudes.

Vigaṇaya, vigaṇayya Having computed.

Vi Decreased. Lit. "is removed".

Vidhi Operation. Method.

Vidhāna Method.

Vinādika Time unit equal to 1/60th of a *nāḍī*.

Viparītakarma The reversed operation.

Vibhāga Partition.

>*Jyāvibhāga* "A partition of chords", see *jyā*.

Vibhājed One should divide.

Virahita Decreased.

Viloma Reversed. Opposite directions. Retrograde.

>*Vilomakarman* A reversed operation. *Vilomagati* is a retrograde motion. *Vilomavivara* is the distance of ⟨two bodies moving in⟩ opposite directions.

Vivara Distance. See *viloma*.

Viśeṣa Difference.

Viśodhayed One should subtract.

Viṣama Uneven. Odd. This word is also used with a different meaning in BAB.1.1, where it is the name given to equations with several unknowns.

>*Viṣamacaturbhuja* "An uneven quadrilateral", i.e. in Bhāskara's commentary a non-isoceles trapezium. However Bhāskara notes in BAB.2.8 that in other treatises this could refer to any quadrilateral.

Viṣkambha Diameter.

>*Viṣkambhārdha* The semi-diameter or radius.

Vistara Width. Bhāskara in BAB.2.8. interprets the word as meaning literally a kind of spreading.

Vistāra Width. Given as a synonym of *āyāma* (length) in BAB.2.8, however in rectangles it is opposed to this very term.

Vṛtta A circle, circular.

>*Vṛttakṣetra* A circular field. Given in BAB.2.9.cd as a synonym of *paridhi*, it then would mean circumference, although it is not used with this meaning in the commentary on the *gaṇitapāda*. *Samavṛttaparidhi* is interpreted by Prabhākara as a *bahuvrīhi*, meaning literally: an evenly-circular circumferenced ⟨field⟩; Bhāskara explains that this interpretation understands the compound as refering to a disk. The same compound is analysed as a *karmadhāraya* by Bhāskara meaning literally: a circumference which is evenly circular.

>In Āryabhaṭa's verses, in the chapter on the sphere (*golapāda*), *vṛtta* is used to characterise the sphericity of three dimensional objects. In BAB.2.7cd *gola* is paraphrased by *vṛtta* in the compound *ghanagolaphala*. In this compound *ghanagola* is a sub-*kamadhāraya* therefore *gola* and with it *vṛtta* means rather "a *circular* solid", rather than "a *sphere/circle* which is a solid".

>*Svavṛtta* is one's own circle. It is the circle having for center the tip of the shadow of a gnomon, whose radius extends to the tip of the gnomon.

Vṛddhi Increase. Common difference in an arithmetically series. Interest in commercial problems. This is a word only used by Bhāskara.

Velā Time.

Velākuṭṭākāra The time-pulverizer.

Vyavahāra Name of a set of eight subjects that form mathematics understood as a global subject (only part of which is presented in the *Āryabhaṭīya*.

Lokavyavahāra "Wordly practice", the particular case where a rule is applied, or the common use of a rule.

Vyāvahāragaṇita Practical computation. Companion term of *sūkṣmagaṇita*, an accurate computation.

Vyākhyāna Explanation. commentary. Used by Bhāskara to characterize his own work in the introductory verse of the chapter on mathematics.

Vyāsa Diameter (literally the seperating ⟨line⟩), *vyāsārdha* is the semi-diameter. This word is given as a synonym of *viṣkambha* in BAB.2.7.ab.

Śa

Śara Arrow. One of the segments of a bow-field, illustrated in Figure 63, page 198.

Śaṅku Gnomon, by extension ⟨the height of⟩ a gnomon; the Rsine of the altitude. For the relation between the size of the gnomon and the Rsine of altitude see the Annex on BAB.2.14.

Śāstra Science, treatise.

Śeṣa Remainder (of a subtraction). Residue.

Maṇḍalaśeṣa is "the residue of revolutions", that is the non-integer part of the number of revolutions performed by a planet since the beginning of the *Kaliyuga* (measured in signs, degrees and minutes).

Rāśiśeṣa is "the residue of signs" that is the non-integer part of the number of signs crossed by a planet since the beginning of the *Kaliyuga* (measured in degrees and minutes).

Bhāgaśeṣa is "the residue of degrees" that is the non-integer part of the number of degrees crossed by a planet since the beginning of the *Kaliyuga* (measured in minutes).

Śṛṅgāṭaka Probably an equilateral, triangular based pyramid, *with* the perpendicular issued from one of its tops onto the triangular base. It is illustrated in Figure 65, page 201.

Śravaṇa Ear, side of a geometrical field.

Średhī Series.

Sa

Saṅkalanā Summation.

Saṅkalanāsaṅkalanā The "summation of a summation", this is the name given by Bhāskara to the sum of the series of progressive sums of natural numbers (i.e. the sum of the series $1, 1+2, 1+2+3, \cdots, 1+2+\cdots+i, \cdots$).

Saṅkhyā Number, amount, value. Calculation.

Saṅkhyāsthānāḥ "The places of numbers", the places in which digits are written in the decimal place-value notation.

Sata Correct ⟨value⟩. Companion term of *asata* (incorrect ⟨value⟩).

Sadṛṣa Same kind. Equal.

Used with the first meaning for the result of the transformation of an integer increased or decreased by a fractional part into a fraction with only a numerator and a denominator. Also used to characterise the type of quantity which enters the multiplication when squaring and cubing.

Sama Same. Equal. Even. Pair. This word does not seem to have exactly the same meaning for Āryabhaṭa and for Bhāskara. For the first, it would have had the meaning "even", in the sense of "uniform"; the meanings understood by the commentator are those given as entries.

Dvisamatryaśrakṣetra, Lit. a three sided field with two equal sides, we have translated it as an "isoceles trilateral".

Samacaturaśrakṣetra, An equi-quadrilateral field, *samacaturaśratā*, lit. the quality of being an equi-quadrilateral; we have translated this expression by "equi-quadrilateralness", *samacaturaśratvā*, the state of being an equi-quadrilateral.

Dvisamacaturbhuja, Lit. a field with four sides, two ⟨of which⟩ are equal, is "an isoceles quadrilateral" i.e. an isoceles trapezium.

Samakaraṇa Lit. making equal. An equation.

Samadalakoṭi Perpendicular. According to Bhāskara, other scholars interpret this word as a *karmadhāraya* meaning a mediator.

Samapariṇāha An even circumference.

This compound is analysed by Bhāsakara as a *karmadhāraya*, meaning literally: that field which is and evenly circular and a circumference (an evenly circular circumference). According to our commentator other scholars interpreted it as a *bahuvrīhi* meaning lit.: that field which has an even circumference (i.e. a disk).

Samavṛttaparidhi See *Vṛtta*.

Samasta Sum. Lit, mingled.

Samāsa Sum. Lit, joining.

Saṃkramaṇa Name of the rule given in Ab.2.24.

Samparka Sum. Vocabulary used by Āryabhaṭa in Ab.2.23 rather than by Bhāskara.

Sampāta ⟨Line whose top is⟩ the intersection. It is a substitute word for *svapātalekha*.

> *Pāta* Means "falling", *sampāta*, "falling together"; this is a substantiated adjective.
>
> In astronomy, this word means "meeting": it is the moment where a planet eclipses another, or the moment of the greatest span of the eclipse.

Saṃyoga, saṃyojamāna Addition.

Saṃvarga Product.

Sahita Increased.

Sūkṣma Accurate. Exact. Companion term of *vyāvahāra* (practical) and of *āsanna* (approximate). Sharp (as the tip of a gnomon), precise.

> *Sūkṣmagaṇita* An accurate computation.

Sūtra Thread or string. It is used in the construction of geometrical figures (as trilaterals and quadrilaterals) and of three dimensional objects (as a gnomon).

> A technical rule given in the form of an aphoristic verse. We have translated it when it is used with the latter meaning as "rule". It can be contrasted with *ārya* and *kārikā* both refering to the verse, in its metrical dimension.

Sthāna Place (for a digit or number).

> *Sthānāntaram* A different place. The next place, to the right or to the left, according to the context, when considering the places in the decimal place-value notation. Maybe in the procedure for extracting the square root, an allusion to a different space where the successive digits of the partial square-root extracted are placed.

Sthāpana Placement. Disposition. Used as an alternative for *nyāsa* "setting-down", which specifies how a quantity or a geometrical field is represented on a working surface.

Sthūlatā The state of being rough.

> *atyantasthūlatā* The state of being exceedingly rough (said of an approximate value).

Sphuṭa Correct, true. Used as a substitute for *sūkṣma* in BAB.2.10.

Svapātalekhā A literal translation would be: "the line on its own falling". This expression names any of the two segments of a perpendicular in a trapezium, as illustrated in Figure 70, page 212. These two segments of the perpendicular (or lines, *lekhā*) are defined from the point of intersection of the diagonals to the middle of the earth and the mouth (the names of the parallel segment in a trapezium). The middle points of the parallel sides being each considered as the "falling" (*pāta*) of the line. However such "lines" are segments of the mediator in isoceles trapeziums but not in uneven trapeziums.

Ha

Hata Multiplied.

Hati Multiplication. Given as a synonym of *saṃvarga* in BAB.2.3ab.

Hīna Decreased.

Hṛta Divided.

Hṛti Division. Given as a synonym of *bhāga* in BAB.2.4.

Hrāsa Subtraction, diminution.

2 Peculiar and metaphoric expressions to name numbers

The reference in parentheses indicates the first occurrence of the expression.

Zero *Kha*, void; *śūnya* (BAB.2.32-33, ex. 14), *viyad*, void (idem, ex. 18), *ākāśa*, idem (idem, ex. 22); *gagana*, the sky (idem, ex. 26).

One *Indu*, the moon (BAB.2.5, ex. 1); *śaśāṅka*, lit. "marked with a rabbit", the moon (BAB.2.32-33; ex. 14); *uḍupa*, the moon (idem, ex. 19); *śītāṃśu*, "with cold rays" i.e. the moon (idem, ex. 20); *śītakiraṇa*, idem (idem, ex. 23); *niśākara*, "the maker of the night", i.e. the moon (idem, ex. 24).

Two *Yama*, a pair (BAB.2.4, ex. 1); *aśvin*, name of the twin sons of the sun (BAB.2.5, ex. 1); *netra* the eyes (BAB.2.32-33, ex. 23); *dasra*, another name of the *aśvins* (idem, ex. 26).

Three *Rāma*, there are three famous Rāmas: the hero of the *Rāmayaṇa*, Balarāma (Kṛṣṇa's brother) and Parasurāma (BAB.2.10, example 2). *Dahana*, fire, as there are three sacrificial fires (BAB.2.11, ex. 1); *hutāśana*, idem; *guṇa* as the three qualities of all created things (truth/goodness for gods (*sattva*), matter/passions for men (*rajas*), darkness/ignorance for demons (*tamas*) (BAB.2.32-33, ex. 19); *śikhin*, fire (idem, ex. 23); *bhuvana* world, as the three worlds of god, men and demons (idem, ex. 24); *puṣkara*, a lake, there are three sacred lakes (idem).

Three and a half *Ardhacaturthā* the fourth ⟨unit⟩ is a half.

Four *Kṛta*, the best of the four casts in a vedic dice game (BAB.2.5, ex.2); *abdhi*, ocean, it is considered that there are four oceans (BAB.2.5, ex.2); *sāgara*, ocean (BAB.2.32-33, ex. 14); *udadhi* idem (idem, ex. 24).

Five *Śara*, as the five arrows of Kāma, the god of love (BAB.2.4; ex.1); *viṣaya*, lit. the objects of the senses (BAB.2.32-33, ex. 13); *bhūta*, the five elements (earth, air, fire, water and stone) (idem, ex.20); *iṣu*, arrow (idem, ex. 24); *artha*, as objects of the senses (idem, ex. 26).

Six *Rasa*, perfume, taste. There are six tastes: *kaṭu* (acrid), *amla* (sour, acid), *madhura* (sweet), *lavaṇa* (saline), *tikta* (bitter) and *kaṣāya* (astringent, fragant); *aṅga*, as the six *Vedāṅgas* (BAB.2.32-33, ex. 23); *ṛtu* a season, there are six seasons (idem).

Seven *Muni*, a sage, there are seven great sages or seers (*ṛṣi*) or maybe the seven stars of the constellation Ursa Major (BAB.2.5, ex. 10); *naga*, "that which does not move", a mountain, there are seven chains of mountains (BAB.2.5, ex.2); *bhūdhara*, "supporting the earth" mountains, (BAB.2.32-33, ex. 14); *adri*, mountains (BAB.2.16); *kṣoṇīdhara*, idem (BAB.2.32-33, ex.

23); *kṣamābhṛt*, a mountain (idem); *adri*, mountain (idem); *svara*, the seven notes that can constitute a *rāga* (idem, ex. 26).

Eight *Vasu*, a class of eight deities (BAB.2.5, ex.1); *nāga* elephant; there are eight elephants symbolising the eight cardinal directions (East, West, South, North, South-east, South-west, North-east, North-west) (BAB.2.32-33, ex. 23).

Nine *Randhra*, orifice; the nine orifices of the human body are: the two eyes, the two nostrils, the mouth , the two ears, the sex, the anus (BAB.2.5, ex. 2); *chidra*, idem (BAB.2.32-33, ex. 24); *nanda* either the nine treasures of Kubera or the nine brother-kings called "Nanda" (idem).

Ten *Paṅkti* a verse with ten syllables in a quarter (BAB.2.9ab, ex.1).

Eleven *Śiva*, as the head of a group of eleven gods called collectively *rudra*[138] (BAB.2.32-33, ex. 13).

Fourteen *Manu* the fourteen successive *manus*, progenitors or sovereigns of the earth mentioned in the *Manusmṛti* 1 63[139]. (BAB.2.9.ab. ex.1).

Sixteen *Aṣṭi* a meter with sixteen syllables per quarter of verse (BAB.2.9.ab, ex. 1)

Eighteen *Dhṛti*, name of a meter with eighteen syllables per quarter of verse (BAB.2.32-33, ex. 14).

Nineteen *Ekonaviṃśati*, twenty minus one.

Twenty-one *Trisapta*, three-(times)-seven (BAB.2.32-33, ex. 9).

Twenty-five *Śarakṛti*, the square of five.

Fifty-nine *Navapañca*, lit. nine-five (BAB.2.32-33, ex. 9).

3   Measure units

3.1   Units of length

Aṅgula Smallest unit of length. Literally an *aṅgula* is a finger or a thumb.

Nṛ Lit. a man. 1 *nṛ* = 96 *aṅgulas*= 4 *hastas*.

Yojana A measure of distance. 1 *yojana* = 800 *nṛ*.

Hasta Lit. a hand or forearm. 24 *aṅgulas* = 1 *hasta*.

---

[138]For further information see [Doniger 1975; glossary, p.351]
[139]See also [Doniger 1975; Glossary, p.347]

Table 12: Units of length

	aṅgula	*hasta*	*nṛ*	*yojana*
aṅgula	1			
hasta	24	1		
nṛ	96	4	1	
yojana	76800	1200	800	1

3.2 Measures of weight

Karṣa 4 *karṣas* = 1 *pala*.

Kuḍuva 1 *kuḍuva* = 4 *setikas*.

Guñjā 5 *guñjās* = 1 *māṣaka*. Used traditionaly by jewelers.

Pala 4 *karṣas* = 1 *pala*. 1 *bhāra* = 2000 *palas*.

Bhāra 1 *bhāra* = 2000 *palas*.

Mānaka 4 *mānakas* = 1 *setikā*.

Māṣaka 5 *guñjās* = 1 *māṣaka*.

Setikā 1 *setikā* = 4 *mānakas*. 4 *setikās* =1 *kuḍuva*.

Sauvarṇika Equal to a *karṣa*? Measure of weight specific to gold.

Table 13: Units of weight

Measures of Grain			
	mānaka	*setikā*	*kuḍuva*
mānaka	1		
setikā	4	1	
kuḍuva	16	4	1

Measures of Gold					
	guñja	*māṣaka*	*karṣa*	*pala*	*bhāra*
guñja	1				
māṣaka	5	1			
karṣa/sauvarṇika	80	16	1		
pala	320	64	4	1	
bhāra	640 000	128 000	8 000	2 000	1

3.3 Coins

One name of a specific coin (*dravya*) is mentioned in the commentary, without any given value: *dīnāra*.

Rūpaka Probably the ancestor of the rupee. 1 *rūpaka* = 20 *vimśopakas*.

Vimśopaka 20 *vimśopakas* = 1 *rūpaka*.

3.4 Time units

Ghaṭikā One sixtieth of a day, half a *muhūrtta* or twenty-four *liptās*. A *ghaṭikā* originally is the name of a clay pot, and by extension became the name of a water pot used in measuring time, and especially the *ghaṭikās* of the day.

Naḍī or nāḍika A synonym of *ghaṭikā*. Half a *muhūrtta*, or 1/60th of a day.

Muhūrtta or Muhurta 1/30th of a day, roughly 48 minutes.

Yāma 1/8th of a day or 3 hours.

Liptā Minute.

Vināḍika 1/60th of a *nāḍī*.

Table 14: Divisions of the day

	dina	*yāma*	*muhurta*	*nāḍika*	*vināḍika*
dina (a day)	1	8	30	60	3600
yāma		1	3 + 3/4	7 + 1/2	450
muhurta or *muhūrtta*			1	2	120
nāḍika, naḍī or *ghaṭikā*				1	60
vināḍika					1

3.5 Subdivisions of a circle

Rāśi A sign. 1/12th of the circumference of the circle.

Liptā A minute. 1/3600th of the circumference.

Kalā A minute. 1/3600th of the circumference.

Bhāga A degree. 1/60th of the circumference.

4 Names of planets, constellations, zodiac signs

The first occurrence of the name is indicated in between parenthesis.

Aśvinī Name of a *nakṣatra-* roughly, a constellation–derived from the names of the twin vedic gods *Aśvin*. Contains stars of what is called today the Taurus constellation.

Balance *Tulādharanara*, litt. the man holding a balance or balance holder (BAB.2.32-33, ex. 14).

Earth *Ku* (Ab.2.1).

Jupiter *Guru*, (Ab.2.1); *adhirūḍhamahendrasūrau* (BAB.2.32-33, ex. 17).

Leo *Mṛgapati*, lord of the beasts (Ab.2.32-33, ex. 7).

Mars *Kuja*, born from the earth (Ab.2.1), *medinīhṛdayaja*, born in the heart of the earth (BAB.2.32-33, ex.16);*aṅgāraka*(BAB.2.32-33, ex. 23); *bhauma* "produced from the earth" (idem).

Mercury *Budha* (Ab.2.1).

Moon *Śaśin*, lit. that which has a rabbit (Ab.2.1), *candra*, lit. that which is bright; *candramas* (BAB.2.32-33, ex.13); *niśānātha*, litt. lord of the night (idem, ex. 14).

Rāhu *tamomaya* "made of darkness" (BAB.2.18, ex. 1).

Saturn *Koṇa*, (Ab.2.1).

Sagittarius *Dhanu*, bow (BAB.2.32-33 ex. 7); *Dhanvin*, the archer (BAB.2.32-33 ex. 12).

Sun *Ravi* (Ab.2.1), *Mayūkhamāla*, litt. wreathed with rays (BAB.2.1); *sahasramarīca*, "with a thousand rays" (BAB.2.16.); *Sūrya* (BAB.2.32-33),*savitṛ* (Ab.2.32-33, ex. 7); *bharttur divasasya, dinabharttur*[140] "lord of the day" (idem, ex. 9); *bhānu* (a ray of light, by extension) (idem, ex.12) *divasakara*, litt. maker of days (idem. ex.13); *arka*, vedic ray of light (idem, ex.14); *bhāsvat* "with lustre" (idem. ex. 19); *tigmāṁśu*, "with harsh rays" (idem, ex.21).

Venus *Bhṛgu* (Ab.2.1).

5 Days of the week

Appear in commentary to verses 32-22

Monday *Somadina* (com. preceding Example 12).

Wednesday (Mercury day) *jñavāra, rātreḥpātustanujadivasa*, litt. the son of the protector of the night (the moon) (ex.12), *budhadivasa* (resolution of ex. 12).

Thursday (Jupiter day) *jīvavāra* (ex.12).

Friday (Venus day) *śukravāra* (ex.12).

Saturday (Saturn day) *śanaiścarasya divasa* (ex.14).

Sunday (sun day) *Sūryadina* (com. preceding Example 12).

---

[140]In classical Sanskrit the word "lord" is usually written with one 't': *bhartur*. This may be the trace of some dialectical writting or just a scribal error.

6 Gods and mythological figures

They do not appear often in the text, however occasionally, in examples, numbers'
names and in the introductory verses, reference are made to some elements of
Hindu Mythology. Therefore, we will briefly give some explanations on this topic.

One thing to bear in mind is that roughly the three major gods of Hinduism
are Brahmā (the creator and grandfather), Śiva (the destroyer) and Viṣṇu (the
preserver). Viṣṇu has eleven incarnations (*avatāra*). Brahmā, a masculine noun (in
the nominative case) is the god; when a neutral noun, *brahman*, it is a philosophical
concept[141]. Āryabhaṭa was a worshiper of Brahmā, a fact quite rare in India today,
Bhāskara was a worshiper of Śiva, as the first verse introducing the *gaṇitapāda*
seems to indicate.

Kṛṣṇa Is the 8th *avatāra* of Viṣṇu.

Brahmā The "Lotus-Born" (*Kamalodbhava*), Brahmā is said to be born from a
lotus growing out of Viṣṇu's navel (BAB.introduction to Ab.2).

The "Creator"(*vedhas*); *Svāyambhū*, litt. self-existent or self-created; gives
the name to the *Svāyambhuvasidhānta*(BAB.2.1).

Ka, lit. "who?", would have arisen from the interpretation of a vedic verse:
'Who (*ka*) knows whence this creation was born?', later interpreted as: '⟨The
god⟩ *ka* knows whence this creation was born.' [142]

Rāhu The demon of eclipses. He is thought to swallow the moon or part of it
during an eclipse.

7 Cardinal directions

North *Uttara*.

South *Dakṣiṇa* (at the right).

East *Pūrva, purastāt* (in front).

West *Apara, paścād* (the last).

---

[141]The essence of all things, the absolute see [Biardeau 1981; p.24-28, and glossaire, p. 183]
[142]See [Doniger 1975; p. 139, note 2]

Bibliography

This Bibliography has two sections: one for Primary sources and the other for Secondary sources. The entries for Primary sources are listed in sub-categories for each Sanskrit author. All entries are then listed according to the roman alphabetical order.

A Primary sources

Āryabhaṭa

[**Clark 1930**] E. C. Clark. *The Āryabhaṭīya of Āryabhaṭa.* University of Chicago Press, 1930.

[**Kern 1874**] H. Kern. *The Āryabhaṭīya, with the commentary Bhaṭadīpikā of Parameśvara.* Reprint from Eastern Book Linkers, 1874.

[**Sengupta 1927**] P. C. Sengupta. The *Aryabhatiyam*, translation. In *Journal of the Department of Letters of the University of Calcutta*, XVI: 1-56, 1927.

[**Sharma & Shukla 1976**] K. V. Sharma and K. S. Shukla. *Āryabhaṭīya of Āryabhaṭa, critically edited with translation.* Indian National Science Academy (INSA), New-Delhi, 1976.

[**Pillai & Sastri 1957**] S. K. Pillai and S. Sastri. *The Āryabhaṭīya with the Bhāṣya of Nīlakaṇṭha Somastuvan.* Trivandrum Sanskrit Series, 101 (1930), 110 (1931) and 185 (1957).

Bakhshālī Manuscript

[**Hayashi 1995**] Takao Hayashi. *The Bhakhshālī Manuscript, An ancient Indian mathematical treatise.* Egbert Forsten. Groningen, 1995.

Brahmagupta

[**Chatterjee 1970**] Bina Chaterjee. *The Khaṇḍakhāyaka of Brahmagupta with the commentary of Bhaṭṭotpala* (2 volumes). Delhi: Motilal, 1970.

[**Dvivedi 1902**] S. Dvivedi. *The Brāhmasphuṭasiddhānta of Brahmagupta.* The Pandit, 1902.

Bhāskara I

[**Apaṭe 1946**] M. C. Apaṭe. *The Laghubhāskarīya, with the commentary of Parameśvara.* Anandāśrama, Skt. series no. 128, Poona, 1946.

[**Sastri 1957**] K. Sastri. *Mahābhāskarīya of Bhāskarācārya with the Bhāṣya of Govindasvāmin and Supercommentary Siddhāntadīpikā of Parameśvara.* Madras Govt. Oriental series, no. cxxx, 1957.

[**Shukla 1960**] K. S. Shukla. *Mahābhāskarīya, Edited and Translated into English, with Explanatory and Critical Notes, and Comments, etc.* Department of mathematics, Lucknow University, 1960.

[**Shukla 1963**] K. S. Shukla. *Laghubhāskarīya, Edited and Translated into English, with Explanatory and Critical Notes, and Comments, etc.,* Department of mathematics and astronomy, Lucknow University, 1963.

[**Shukla 1976**] K. S. Shukla. *Āryabhaṭīya of Āryabhaṭa, with the commentary of Bhāskara I and Someśvara.* Indian National Science Academy (INSA), New-Delhi, 1976.

Varāhamihira

[**Neugebauer & Pingree 1971**] Otto Neugebauer and David Pingree. *The Pañcasiddhāntikā of Varāhamihira.* Kopenhagen, 1971.

B   Secondary sources

[**Ahmad 1981**] Afzal Ahmad. 'On the π of Āryabhaṭa I' in *Gaṇita Bhārati* 3 (3-4): 83-85, 1981.

[**Apte 1957**] V.S. Apte. *The practical Sanskrit-English Dictionary.* Poona, 1957.

[**Bag 1976**] A. K. Bag. *Mathematics in Ancient and Medieval India.* Chukhambha Orientala, 1976.

[**Bhattacharya 1991**] Ramkrishna Bhattacharya. 'The case of Āryabhaṭa and his detractors' in *The Indian Historical Review*, XVII (1-2: 35-47, July 1990-January 1991.

[**Biardeau 1981**] Madeleine Biardeau. *L'Hindouisme, Anthropologie d'une civilisation.* Flammarion, 1981.

[**Billard 1971**] André Billard. *L'Astronomie Indienne*, EFEO, 1971.

[**Billard 1977**] André Billard. 'Āryabhaṭa and Indian astronomy' in *Indian Journal of History of Science (IJHS)*, 12(2): 207-224, 1977.

[**Bronkhorst 1990**] Johannes Bronkhorst. 'Vārttika' in *Wiener Zeitschrift fuer die Kunde Suedasiens*, 34:123-146, 1990.

[**Bronkhorst 1991**] Johannes Bronkhorst. 'Two literary conventions of Classical India' in *Asiatische Studien/ Études Asiatiques*, 45 (2): 210-227, 1991.

[**Bryant, 2001**] Edwin Bryant. *The Quest for the Origins of Vedic Culture: The Indo-Aryan Migration Debate.* Oxford University Press. 2001.

[**CESS**] *Census of the Exact Sciences in Sanskrit, series A, 5 volumes*, David Pingree. American Philosophical Society, 1970-1995.

[**Chemla 1992**] Karine Chemla. 'Résonances entre démonstration et procédure. Remarques sur le commentaire de Liu Hui (IIIème siècle) aux *Neuf Chapitres sur les Procédures Mathématiques* (Ier siècle)' in *Extrême Orient, Extrême Occident*, 14: 91-129, 1992.

[**Chemla 1994**] Karine Chemla. 'Similarities between Chinese and Arabic mathematical writings: (I) root extractions' in *Arabic Sciences and Philosophy*, 4: 207-266, 1994.

[**Chemla 1995**] Karine Chemla. 'Histoire des Sciences et Matérialité des textes, proposition d'enquête' in *Enquête*, 1: 167-180, 1995.

[**Chemla 1996**] Karine Chemla.'What is the content of this book? a plea for developing History of Science and History of Text conjointly' in *Philosophy and the History of Science*, 1996.

[**Chemla 1997**] Karine Chemla. 'Qu'est-ce qu'un problème dans la tradition mathématique de la Chine ancienne?' in *Extrême Orient, Extrême Occident*, 19: 91-126, 1997.

[**Chemla & Keller 2002**] Karine Chemla and Agathe Keller. 'The Chinese Mian and the Sanskrit *karaṇīs*' in *From China to Paris: 2000 Years of Mathematical Transmission*, Dauben, Dold and Van Dallen, eds. Steiner Verlag, Stuttgart, RFA, pp. 87-132, 2002.

[**Coward & Raja 1990**] Harold G. Coward and K. Kunjuni Raja. *Encyclopedia of Indian Philosophies, volume V: The Philosophy of the Grammarians.* Motilal Banarsidas, 1990.

[**Datta & Singh 1938**] B. Datta and S. N. Singh. *History of Hindu mathematics*, Asia Publishing House, 1938.

[**Datta & Singh 1980**] B. Datta and S. N. Singh ; revised by K. S. Shukla 'Hindu geometry' in *Indian Journal of History of Science*, 15 (2): 121-187, 1980.

[**Datta & Singh 1983**] B. Datta and S. N. Singh ; revised by K. S. Shukla. 'Hindu trigonometry' in *Indian Journal of History of Science*, 18 (1): 309-108, 1983.

[**Datta 1926**] B. Datta. 'Hindu values of π' in *Journal of the Asiatic Society of Bengal*, New series, 22:25-42, 1926.

[**Datta 1930**] B. Datta. The two Bhāskaras. *Indian Historical Quarterly*, VI (4): 727-736, 1930.

[**Datta 1932**] B. Datta. 'The Elder Āryabhaṭa's rule for the solution of indeterminate equations of the first degree' in *Bulletin of the Calcutta Mathematical Society*, XVIV (1): 19-36, 1932.

[**Doniger 1975**] Wendy Doniger O'Flaherty. *Hindu Myths*. Penguin, 1975.

[**Filliozat & Renou 1949**] Jean Filliozat et Louis Renou. *L'Inde Classique, manuel des études indiennes*. Payot, 1949.

[**Filliozat 1988a**] Pierre-Sylvain Filliozat. 'Calculs de demi-cordes d'arcs par Āryabhaṭa et Bhāskara I' in *Bulletin des Études Indiennes*, 6: 255-274, 1988.

[**Filliozat 1988b**] Pierre-Sylvain Filliozat. *Grammaire Sanskrite Pāṇinéenne*. Picard, 1988

[**Filliozat 1990**] Pierre-Sylvain Filliozat. 'Yukti le quatrième *pramāṇa* des médecins' in *Journal of the European Ayurvedic Society*, 1: 33-46, 1990.

[**Filliozat & Mazars 1985**] P. -S. Filliozat and Guy Mazars. 'Observations sur la formule du volume de la Pyramide et de la Sphère chez Āryabhaṭa' in *Bulletin des Études Indiennes*, 3:37-48, 1985]

[**Ganguli 1932**] S. Ganguli. 'India's contribution to the Theory of Indeterminate Equations of the First Degree' in *Journal of the Indian Mathematical Society / New Quarterly*, 19:110-168, 1931/32.

[**Gareth & Jones 1998**] J. Mary Gareth and A. Jones. *Elementary Number Theory*. Springer, 1998.

[**Gupta 1977**] R. C. Gupta. 'On some mathematical rules from the *Āryabhaṭīya*' in *Indian Journal of History of Science*, 12 (2): 200-206, 1977.

[**Hayashi 1977**] Takao Hayashi. '*Karaṇī* and the *karaṇī*-operation' in *Japanese Studies in the History of Science*, 17, pp. 51-59. 1977.

[**Hayashi 1994**] Takao Hayashi. 'Indian mathematics' in *Companion Encyclopedia of History and Philosophy of the Mathematical Sciences*. volume 1: pp. 118-130. Edited by I. Grattan-Guiness, Routledge, London, 1994.

[**Hayashi 1997a**] Takao Hayashi. 'Āryabhaṭa's rule and table for sine-differences' in *Historia Mathematica*, 24: 396-406, 1997.

[**Hayashi 1997b**] Takao Hayashi. 'Calculations of the Surface of a sphere in India' in *The Science and Engineering Review of Doshisha University*, 37 (4): 194-238, 1997.

[**Hayashi, Kusuba & Yano 1989**] Takao Hayashi, Takanori Kusuba and Mishio Yano. 'Indian values of π derived from Āryabhaṭa's value' in *Historia Scientarium*, 42: 33-44, 1989.

[**Houben 1997**] Jan E. M. Houben. '*Sūtra* and *bhāṣyasūtra* in Bhartṛhari's Mahā-bhāṣya dīpikā: on the theory and practice of a scientific and philosophical genre' in *India and Beyond, Aspects of Literature, Meaning, Ritual and Thought, Essays in honor of Frits Staal*, Dick van der Meij editor, Kegan Paul International, Chapter XV, pp. 271-305, 1997.

[**Jain 1995**] Pushpa Kumari Jain. *A Critical edition, English translation and commentary of the Upodghāta Ṣaḍvidhaprakaraṇa and Kuṭṭakādhikāra of the Sūryaprakāsa of Sūryadāsa.* PhD Thesis, Simon Fraser University, 1995.

[**Kale 1972**] M. R. Kale. *A Higher Sanskrit Grammar.* Motilal Banarsidas, 1972.

[**Kaye 1908**] G. R. Kaye. 'Notes on Indian mathematics. no. 2. -Āryabhaṭa' in *Journal of the Asiatic Society of Bengal*, IV (3): 111-141, 1908.

[**Keller 1995**] Agathe Keller. *Fractions et Règles de Trois en Inde au Vèmes et VIIèmes siècles, deux vers de l'Āryabhaṭīya d'Āryabhaṭa, accompagnés du commentaire de Bhāskara.* Mémoire de DEA, Université Paris VII, 1995.

[**Keller 2000**] Agathe Keller. *Un commentaire indien du VIIe siècle: Bhāskara et le* gaṇitapāda *de l* Āryabhaṭīya. Thèse de l'Université Paris VII.

[**Keller 2005**] Agathe Keller. *Making Diagrams Speak, in Bhāskara I's commentary on the Āryabhaṭīya.* Historia Mathematica 32: 275-302.

[**Keller forthcoming**] Agathe Keller. 'Qu'est-ce que les mathématiques ? les réponses taxinomiques de Bhāskara un commentateur, mathématicien et astronome indien du VIIème siècle'. Actes du Colloque Des sciences, des frontières, Université de Nancy, France.

[**Kusuba 1981**] 'Brahmagupta's Sutras on Tri- and Quadrilaterals.' *Historia Scientiarum* 21, 43–55, 1981.

[**Kusuba 1987a**] 'Brahmagupta's *Pāṭīgaṇita*.' *Medieval mathematics*, ed. by S. Ito, Tokyo: Kyoritsu Shuppan, pp. 381–407, 1987 (in Japanese).

[**Kusuba 1987b**] 'Brahmagupta's *Bījagaṇita*.' *Medieval mathematics*, ed. by S. Ito, Tokyo: Kyoritsu Shuppan, pp. 408–428 , 1987 (in Japanese).

[**Kusuba 1987c**] 'India in the History of Science.' *Comparative Studies of History of Science*, History of Science Course, ed. by S. Ito and Y. Murakami, vol. 3, Tokyo: Baihūkan, pp. 105–128, 1987 (in Japanese).

[**Kusuba & Pingree 2002**] *Arabic astronomy in Sanskrit*. Islamic Philosophy, Theology and Science vol. 47. Brill 2002. (in collaboration with David Pingree)

[**Maiti 1994**] N. L. Maiti. 'Notes on Broken Bamboo Problems' in *gaṇita Bhāratī*, 16 (1-4):25-36, 1994.

[**Maiti 1996**] N. L. Maiti. 'Antiquity of Trairāśika in India' in *gaṇita Bhāratī*, 18 (1-4): 1-8, 1996.

[**Markel 1995**] Stephen Markel. *Origins of the Indian planetary Deities*. Studies in Asian Thought and Religion. volume 16. Edwin Mellen Press, 1995.

[**Mishra & Singh 1996**] V. Mishra and S. L. Singh. 'Height and distance Problems in Ancient Indian mathematics' in *Gaṇita Bhāratī*, 18 (1-4): 25-30, 1996.

[**Ōhashi 1994**] Yukio Ōhashi. 'Astronomical Instruments in Classical Siddhāntas' in *Indian Journal of History of Science*, 29 (2): 155-313, 1994.

[**Pingree 1981**] David Pingree. *JYOTIḤSĀSTRA, Astral and Mathematical Literature*. Otto Harowitz, 1981.

[**Pingree 1993**] David Pingree. 'Āryabhaṭa, the *Paitāmahasiddhānta*, and Greek astronomy' in *Studies in the History of Medecine & Science, New series* XII (1-2): 69-79, 1993.

[**Renou 1963**] Louis Renou. 'Sur le genre du sūtra dans la littérature Sanskrite' in *Journal Asiatique*, 251, 1963.

[**Renou 1984**] Louis Renou. *Grammaire Sanskrite*. J. Maisonneuve, 1984.

[**Sarasvati 1979**] T. A. Sarasvati Amma. *Geometry in Ancient and Medieval India*. Motilal Banarsidas, 1979.

[**Sarma 1977**] K. V. Sarma. *A History of the Kerala School of Hindu Astronomy*, Viveshvarand Institute Publications, Hoshiarpur, 1972.

[**Sarma 1977**] K. V. Sarma. 'Tradition of *Āryabhaṭīya* in Kerala: revision of planetary parameters' in *Indian Journal of History of Science*, 12 (2): 194-199, 1972.

[**Sarma 2002**] S. R. Sarma 'Rule of Three and its Variations in India' in *From China to Paris: 2000 Years of Mathematical Transmission*, Steiner Verlag, Stuttgart, RFA, pp. 133-156, 2002.

[**Shukla 1971 a**] K. S. Shukla. 'Hindu mathematics in the Seventh Century as found in Bhāskara I's commentary of the *Āryabhaṭīya* (I)' in *gaṇita*, 22 (1): 115-130, 1971.

[**Shukla 1971 b**] K. S. Shukla. 'Hindu mathematics in the Seventh Century as found in Bhāskara I's commentary of the *Āryabhaṭīya* (II)' in *gaṇita*, 22 (2): 61-78, 1971.

[**Shukla 1972a**] K. S. Shukla. 'Hindu mathematics in the Seventh Century as found in Bhāskara I's commentary of the *Āryabhaṭīya* (III)' in *gaṇita*, 23 (1): 57-79, 1972.

[**Shukla 1972b**] K. S. Shukla. 'Hindu mathematics in the Seventh Century as found in Bhāskara I's commentary of the *Āryabhaṭīya* (IV)' in *gaṇita*, 23 (2): 41-50, 1972.

[**Srinivas 1990**] M. D. Srinivas. 'The methodology of Indian mathematics and its contemporary relevance' in *History of Science and Technology in India; Vol. I, Mathematics and Astronomy*, sundeep Prakashan, pp. 28-86, 1990.

[**Thapar 1966**] Romila Thapar. *A History of India, volume I.* Penguin, 1966.

[**Yano 1977**] Michio Yano. 'Three types of Hindu sine Tables' in *Indian Journal of History of Science*, 12 (2): 83-89, 1977.

[**Yano 1978**] Michio Yano. 'Someśvara on the Āryabhaṭīya' in *Indian and Buddhist studies*, XXVI (II): 38-41, 1978.

[**Yano 1980**] Michio Yano. 'Āryabhaṭa's possible rebuttal to objections to his Theory of the Rotation of the Earth' in *Historia Scientarum*, 19: 101-105, 1980.

Index